WOI

Pocket Guide
to
Mobile Connectivity

2nd Edition

Telecommunications
Cell Communications
Internet Connectivity
Mobile Connectivity

World Trade Press

Pocket Guide to Mobile Connectivity

2nd Edition

Telecommunications
Cell Communications
Internet Connectivity
Mobile Connectivity

Sibylla Putzi-Ortiz • Myron Manley
Wendy Bidwell • Paul Denegri
—
Gilbert Chamaa • Brian Duffey • Gary Fox
Nicolette Dalpino • Edward Hinkelman
Blythe Millington • Jason Mann
Rattany Khiev • Emily Hansen

WORLD TRADE PRESS®
Books, E-Content & Maps for International Trade

Pocket Guide to Mobile Connectivity, 2nd Edition
ISBN 1-885073-71-2

Publisher
World Trade Press
1450 Grant Avenue, Suite 204
Novato, California 94945 USA
Tel: [1] (415) 898-1124
Fax: [1] (415) 898-1080
USA order Line: [1] (800) 833-8586
E-mail for individual orders: orders@worldtradepress.com
E-mail for bulk orders: sales@worldtradepress.com
www.worldtradepress.com
www.globalroadwarrior.com
www.howtoconnect.com
www.worldtraderef.com
www.bestcountryreports.com
www.atlascartographic.com

Production Credits
Cover design: Cyndie Wooley and Lise Stampfli
Electric plug and phone plug illustrations: Gary Fox
Cell and satellite phone and Internet illustrations: Jason Mann, www.manndesignstudios.com
Desktop publishing: Sibylla Putzi, Edward Hinkelman

Disclaimer
Material for this publication has been obtained from agencies of various countries, embassies, consulates, and from interviews and correspondence. We have diligently tried to ensure the accuracy of all the information and to present as comprehensive a reference work as space would permit. If we find errors we strive to correct them in preparing future editions. The publishers, however, take no responsibility for inaccurate or incomplete information that may have been submitted to them in the course of research for this publication. The facts published indicate the result of those inquiries and no warranty as to their accuracy is given.

Library of Congress Cataloging-in-Publication Data
Pocket Guide to Mobile Connectivity: telecommunications, cell communications, Internet connectivity, mobile connectivity -- 2nd ed.
p. cm.
2005 ISBN 1-885073-71-2
1. Mobile communication systems--Handbooks, manuals, etc.
2. Computer networks--Remote access--Handbooks, manuals, etc.
3. Mobile computing--Handbooks, manuals, etc.
4. Wireless Internet--Handbooks, manuals, etc.
5. Cellular telephone systems--Handbooks, manuals, etc.
I. Title: World Trade Press Pocket Guide to Mobile Connectivity.
TK6570.M6P63 2005
384--dc22   2005048994
Printed in the United States of America

Table of Contents

Personal Technical Data Page

☛ **CAUTION**: Enter password and other important data only if you are prepared to carefully protect the custody of this book at all times.

Cell Phone
Cell Phone #: _____

Make / Model: _____

Serial #: _____

Domestic Service Provider: _____

Support Contact Telephone #: _____

Laptop Computer
Make / Model: _____

Serial #: _____

Modem Type / Speed: _____

Purchased From: _____

Support Contact Telephone #: _____

E-Mail Software
Make / Release #: _____

Serial #: _____

Support Contact Telephone #: _____

ISP (Internet Service Provider)
ISP Name: _____

Dial-up Access #: _____

ISP Support Telephone # _____

Data Software
Make / Release #: _____

Serial #: _____

Support Contact Telephone #: _____

Modem Software
Make / Release #: _____

Serial #: _____

Support Contact Telephone #: _____

FTP (File Transfer Protocol) Data
Host Name / IP #: _____

Username: _____

Password: _____

General Data / Dialing Information
Office Technical Support Person: _____

Office Support Person Telephone #: _____

Office Support Person E-Mail: _____

Notes

Save Money!

13 Ways to Save Money
on International Communications

1. **Don't Use Your Hotel Telephone.** Hotels are notorious for excessive charges on long-distance and international calls. Use a cell phone, calling card or public telephone when possible.

2. **Make Your Calls at Off-Peak Hours.** Almost all telecommunications carriers have lower rates in the evenings and on the weekends.

3. **Be Purposeful in Your Calls.** Make your conversations short and to the point. Instruct your staff that you will call at a particular time each day and that you want certain information (messages, sales figures, deposits) ready.

4. **Write Your E-Mail Messages Off-Line**. When you connect to the Internet download all your messages, disconnect, write your replies and then reconnect only long enough to send your messages.

5. **Don't Do Mindless Web Surfing.** In some countries, especially the U.S., local phone rates and Internet access are inexpensive. An hour of Web surfing might cost US$1. In another country, however, an hour of telephone and Web connectivity might cost US$150.

6. **Use a Prepaid Calling Card.** Almost all major long-distance telecommunications providers offer special calling card rates. These can be as little as 25 percent of the standard local charges.

7. **Use Your Existing Cellular Service Provider.** In many cases your existing cellular provider will have service where you are traveling. Ask them for a service plan that includes airtime in these countries.

8. **Rent a Local Cell Phone.** A cell phone can save money in many ways. The greatest savings is time. Many businesspeople waste huge amounts of time simply locating public telephones.

9. **Use Prepaid SIM Cards.** When using a GSM cell phone, purchasing prepaid SIM cards offers the most inexpensive per-minute charges. See "Cell Phone SIM Cards" on page 32.

10. **Use Text Messaging When Possible.** When using a GSM cell phone, try sending a short text message instead of a voice call whenever the communication is appropriate. Text messages are by far the most inexpensive way to communicate by cell phone. See "SMS & MMS" on page 47.

11. **Connect to Your Home ISP.** It is sometimes less expensive to connect to the Internet when overseas by connecting to your home country ISP! If you use a calling card or a callback service, the cost of the call is not much different than making a local call. Furthermore, you are already paying for your home ISP, so you will incur no additional charges for the Internet connection itself. Of course, it's best to check rates before making a decision. However, establishing a long-distance connection from a third-world country that is good enough for data may be a problem at times.

12. **Consider Getting an Account with VoIP Commercial Services** such as Net2Phone, Dialpad, etc. for very low international rates, even without the use of a computer.

13. **Use a Callback Service.** Callback services can potentially save international travelers a great deal of money. Here is how it works. Phone rates vary from country to country. The rates in the U.S., for example, are much less than in Brazil. A callback user in Brazil calls a local number in Brazil that connects to a number in the U.S. that automatically calls the Brazilian number back and makes the new connection. Hence "call-back." The user is charged a premium on the U.S. rate, but much less than the Brazilian rate. Callback services are understandably met with resistance by monopoly telecom providers. Some governments have passed laws prohibiting callback services. Still, they remain in business.

International Dialing Guide

This guide provides basic data for calling 220 countries and territories around the world. It has been designed to be useful, regardless of the country of origin or country of destination of your call.

How to Dial International Calls

International direct dialing from most countries is quite easy. An international call simply consists of dialing a sequence of numbers as follows:

1. The International Access Code (IAC),

Note: In the case of cell phones, you merely enter the '+' symbol (usually by pressing and holding zero), and the cellular service provider will automatically replace it with the appropriate IAC

2. The destination country code,
3. The city/area code, and
4. The local telephone number.

Each step routes your call a step closer to the person or business you are calling.

The International Access Code (IAC). The IAC prefix is used to get an international line. For example, If you are calling from the United States, the IAC is 011. The IAC differs from country to country. For Brazil it is 00. Some countries do not allow direct dialing for international calls and require an operator. The table on the next page lists IAC prefixes for most countries. You may need to wait for a dial tone after dialing the IAC.

The Country Code. Next comes the code of the country you are calling. Refer to "International Access and Dialing Codes by Country" on page 11, or individual country listings.

The Area Code. These are regional or city dialing codes. Refer to "International Access and Dialing Codes by Country" on page 11 and individual country listings. Some countries have done away with area codes, incorporating them into "national" subscriber numbers.

The Local Telephone Number. This is also called the subscriber number. In some countries, a number may begin with a different digit depending on whether one is dialing a regular landline phone or a cellular phone. In Greece, for example, all landline numbers begin with '2' while cell phone numbers begin with '6'.

Points of Confusion

1. When you get someone's phone number you may also get the IAC, country code and area code along with the actual subscriber number. This can cause confusion, especially when they give you *their* country's IAC rather than *your* IAC.

2. Many business cards have the country's long-distance prefix added to the number. This prefix is used _only_ when making long-distance calls within that country. For example, 0 is the long-distance direct-dial prefix for Germany. This prefix is omitted from an international call to Germany. For example, a number on a business card might read: (030) 8305-0 or (0)30 8305-0. If dialing that number from the U.S. or from anywhere else outside Germany, the caller would dial: their IAC + [49] + (30) + 8305-0.

3. Some countries use the same country code. For example, the country code for the USA, Canada and some Caribbean nations is "1". Calls between these countries are treated simply as a long distance call. In this example you do not use the IAC, but use "1" as the long distance prefix.

Dialing Examples

Example 1

A call from London, U.K. ➜ Chicago, USA:

00 + [1] + (312) + (local number)

Explanation: The UK's IAC is 00, the U.S.' country code is [1] and Chicago's area code is (312).

Example 2

A call from Denmark ➜ Sydney, Australia:

00 + [61] + (2) + (local number)

Explanation: Denmark's IAC is 00, Australia's country code is [61] and Sydney's area code is (2).

Example 3

A call from the USA ➜ Vancouver, BC, Canada:

1 + (604) + (the local number)

Explanation: Both the U.S. and Canada are part of the North American Numbering Plan. The prefix 1 is used for long-distance calls in Canada, the U.S., and the Caribbean. Vancouver's area code is (604).

Example 4

A call from Hong Kong ➜ Delhi, India:

001 + [91] + (11) + (local number).

Explanation: Hong Kong's IAC is 001, India's country code is [91], and Delhi's city code is (11).

Example 5

A call from the USA ➜ Hong Kong:

011 + [852] + (local number)

Explanation: The U.S.'s IAC is 011, Hong Kong's country code is [852], and there are no city/area codes in Hong Kong.

For Further Information

The information in this section is current as of October 2005. Note that dialing codes and dialing systems are changing rapidly as the demand for telephone, fax, moden and data lines increase.

For codes not listed here, or for the current time anywhere in the world, call your international operator. In the United States, dial 00 for AT&T information.

On the Internet, world time by country is at www.globaltimeclock.com.

World time and dialing codes are at www.howtoconnect.html.

Key to International Access and Dialing Codes

✪ Capital city

†† More than one time zone in this country.

* City/area codes not required in this country.

** This city/area code used for entire country or territory.

International Access and Dialing Codes by Country

Country	Country Code	International Access Codes	Capital City City/Area Code
Afghanistan	[93]	00	✪Kabul (20 to 25)
Albania	[355]	00	✪Tirana (4)
Algeria	[213]	00	✪Algiers (21)
American Samoa	[684]	011	✪Pago Pago*
Andorra	[376]	00	✪Andorra la Vella*
Angola	[244]	00	✪Luanda (2)
Anguilla	[1]	011	✪The Valley (264)**
Antigua & Barbuda	[1]	011	✪St. John's (268)**
Argentina††	[54]	00	✪Buenos Aires (11)
Armenia	[374]	00	✪Yerevan (1)
Aruba	[297]	00	✪Oranjestad (8)**
Australia††	[61]	0011	✪Canberra (2)
Austria	[43]	00	✪Vienna (1)
Azerbaijan	[994]	8*10	✪Baku (12)
Bahamas	[1]	011	✪Nassau (242)**
Bahrain	[973]	00	✪Manama *
Bangladesh	[880]	00	✪Dhaka (2)
Barbados	[1]	011	✪Bridgetown (246)**
Belarus	[375]	8*10	✪Minsk (17)
Belgium	[32]	00	✪Brussels (2)
Belize	[501]	00	✪Belmopan (8)
Benin	[229]	00	✪Porto-Novo *
Bermuda	[1]	011	✪Hamilton (441)**
Bhutan	[975]	00	✪Thimphu*
Bolivia	[591]	001 + one-digit carrier code	✪La Paz (2) ✪Sucre (4)
Bosnia & Herzegovina	[387]	00	✪Sarajevo (33)
Botswana	[267]	00	✪Gaborone
Brazil††	[55]	00 + two-digit carrier code	✪Brasilia (61)
Brunei	[673]	00	✪Bandar Seri Begawan (2)

Country	Country Code	International Access Codes	Capital City City/Area Code
Bulgaria	[359]	00	✪Sofia (2)
Burkina Faso	[226]	00	✪Ouagadougou*
Burundi	[257]	90	✪Bujumbura (2)
Cambodia	[855]	00	✪Phnom Penh (23)
Cameroon	[237]	00	✪Yaoundé*
Canada	[1]	011	✪Ottawa, ON (613)
Cape Verde	[238]	0	✪Praia*
Cayman Islands	[1]	011	✪George Town (345)**
Central African Republic	[236]	19	✪Bangui (all six-digit numbers)
Chad	[235]	15	✪N'Djamena*
Chagos Archipelago	[246]	00	✪Diego Garcia*
Chechnya	[7]	8*10	✪Grozny (8712)
Chile	[56]	00 or carrier code + 0	✪Santiago (2)
China (PRC)	[86]	00	✪Beijing (10)
Colombia	[57]	005, 007, or 009	✪Bogota (1)
Comoros	[269]	00	✪Moroni*
Congo, Dem. Rep. of	[243]	00	✪Kinshasa (12)
Congo, Rep. of	[242]	00	✪Brazzaville*
Cook Islands	[682]	00	✪Avarua*
Costa Rica	[506]	00	✪San José*
Côte d'Ivoire	[225]	00	✪Yamoussoukro*
Croatia	[385]	00	✪Zagreb (1)
Cuba	[53]	119	✪Havana (7)
Cuba (Guantanamo Bay)	[5399]	00	U.S. Naval Base
Cyprus	[357]	00	✪Nicosia*
Czech Republic	[420]	00	✪Prague (2)
Denmark	[45]	00	✪Copenhagen*

Key to International Access and Dialing Codes

✪ Capital city
†† More than one time zone in this country.
* City/area codes not required in this country.
** This city/area code used for entire country or territory.

Country	Country Code	International Access Codes	Capital City City/Area Code
Djibouti	[253]	00	✪Djibouti*
Dominica	[1]	011	✪Roseau (767)**
Dominican Republic	[1]	011	✪Santo Domingo (809)**
Ecuador	[593]	00	✪Quito (2)
Egypt	[20]	00	✪Cairo (2)
El Salvador	[503]	00	✪San Salvador*
Equatorial Guinea	[240]	00	✪Malabo (9)
Eritrea	[291]	00	✪Asmara (1)
Estonia	[372]	00	✪Tallinn*
Ethiopia	[251]	00	✪Addis Ababa (1)
Faeroe Islands	[298]	00	✪Tórshavn*
Falkland Islands	[500]	00	✪Stanley* (all five-digit numbers)
Fiji	[679]	00	✪Suva*
Finland	[358]	00	✪Helsinki (9)
France	[33]	00	✪Paris*
French Antilles	[590]	00	✪Guadeloupe*
French Guiana	[594]	00	✪Cayenne*
French Polynesia††	[689]	00	✪Papeete, Tahiti*
Gabon	[241]	00	✪Libreville*
Gambia, The	[220]	00	✪Banjul*
Georgia	[995]	8*10	✪Tbilisi (32)
Germany	[49]	00	✪Berlin (30)
Ghana	[233]	00	✪Accra (21)
Gibraltar	[350]	00	✪Gibraltar*
Greece	[30]	00	✪Athens* (all ten-digit numbers; city codes are now incorporated into local numbers).
Greenland††	[299]	99	✪Nuuk (Godthaab)*
Grenada	[1]	011	✪St. George's (473)**
Guadeloupe	[590]	00	✪Basse-Terre (81)
Guam	[1]	011	✪Agana (671)*
Guatemala	[502]	00	✪Guatemala City*
Guinea	[224]	00	✪Conakry*

Country	Country Code	International Access Codes	Capital City City/Area Code
Guinea-Bissau	[245]	00	✪Bissau* (all six-digit numbers)
Guyana	[592]	001	✪Georgetown*
Haiti	[509]	00	✪Port-au-Prince* (all seven-digit numbers)
Honduras	[504]	00	✪Tegucigalpa (all seven-digit numbers)
Hong Kong	[852]	001	✪Hong Kong
Hungary	[36]	00	✪Budapest (1)
Iceland	[354]	00	✪Reykjavik (all seven-digit numbers)
India	[91]	00	✪New Delhi (11)
Indonesia††	[62]	001 or 008	✪Jakarta (21)
Iran	[98]	00	✪Tehran (21)
Iraq	[964]	00	✪Baghdad (1)
Ireland	[353]	00	✪Dublin (1)
Israel	[972]	00	✪Jerusalem (2)
Italy	[39]	00	✪Rome (06)
Jamaica	[1]	011	✪Kingston (876)**
Japan	[81]	001	✪Tokyo (3)
Jordan	[962]	00	✪Amman (6)
Kazakhstan	[7]	8*10	✪Alma-Ata (3272)
Kenya	[254]	000	✪Nairobi (20)
Korea, North	[850]	00	✪Pyongyang (2)
Korea, South	[82]	001	✪Seoul (2)
Kosovo	[381]	99	✪Pristina (38)
Kuwait	[965]	00	✪Kuwait*
Kyrgyzstan	[996]	00	✪Bishkek (312)
Laos	[856]	00	✪Vientiane (21)
Latvia	[371]	00	✪Riga (no area code used for Riga)

Key to International Access and Dialing Codes

✪ Capital city
†† More than one time zone in this country.
* City/area codes not required in this country.
** This city/area code used for entire country or territory.

Country	Country Code	International Access Codes	Capital City City/Area Code
Lebanon	[961]	00	✪Beirut (1)
Lesotho	[266]	00	✪Maseru*
Liberia	[231]	00	✪Monrovia*
Libya	[218]	00	✪Tripoli (21)
Liechtenstein	[423]	00	✪Vaduz*
Lithuania	[370]	00	✪Vilnius (2)
Luxembourg	[352]	00	✪Luxembourg*
Macau	[853]	00	✪Macau*
Macedonia	[389]	00	✪Skopje (2)
Madagascar	[261]	00	✪Antananarivo*
Malawi	[265]	101	✪Lilongwe* (all six-digit numbers)
Malaysia	[60]	00	✪Kuala Lumpur (3)
Maldives	[960]	00	✪Malé*
Mali	[223]	00	✪Bamako *
Malta	[356]	00	✪Valletta*
Marshall Islands	[692]	011	✪Majuro* Ebeye*
Martinique	[596]	00	✪Fort-De-France
Mauritania	[222]	00	✪Nouakchott*
Mauritius	[230]	00	✪Port Louis*
Mexico††	[52]	00	✪Mexico City (55)
Midway Islands	[808]	011	Sand Island (main island)
Moldova	[373]	8*10	✪Chisinau (2)
Monaco	[377]	00	✪Monaco*
Mongolia	[976]	001	✪Ulaanbaatar (1)
Montserrat	[1]	011	✪Plymouth (664)**
Morocco	[212]	00	✪Rabat (3)
Mozambique	[258]	00	✪Maputo (1)
Myanmar (Burma)	[95]	0	✪Rangoon (1)
Namibia	[264]	09	✪Windhoek (61)
Nepal	[977]	00	✪Kathmandu (1)
Netherlands	[31]	00	✪Amsterdam (20) ✪The Hague (70)

Country	Country Code	International Access Codes	Capital City City/Area Code
Netherlands Antilles	[599]	00	✪Willemstad (9) St. Maarten (5)
New Caledonia	[687]	00	✪Nouméa*
New Zealand	[64]	00	✪Wellington (4)
Nicaragua	[505]	00	✪Managua (2)
Niger Republic	[227]	00	✪Niamey*
Nigeria	[234]	009	✪Abuja (9), ✪Lagos (1)
Niue Islands	[683]	00	✪Alofi*
Northern Mariana Islands	[1]	011	✪Saipan (670)
Norway	[47]	00	✪Oslo* (all eight-digit numbers)
Oman	[968]	00	✪Muscat*
Pakistan	[92]	00	✪Islamabad (51)
Palau	[680]	011	✪Koror*
Panama	[507]	00	✪Panama City* (all seven-digit numbers)
Papua New Guinea	[675]	05	✪Port Moresby*
Paraguay	[595]	00	✪Asuncion (21)
Peru	[51]	00	✪Lima (1)
Philippines	[63]	00	✪Manila (2)
Poland	[48]	0*0	✪Warsaw (22)
Portugal	[351]	00	✪Lisbon* (all nine-digit numbers)
Puerto Rico	[1]	011	✪San Juan (787)**
Qatar	[974]	0	✪Doha*
Reunion	[262]	00	✪St. Denis*
Romania	[40]	00	✪Bucharest (21)
Russia††	[7]	8*10	✪Moscow (095)
Rwanda	[250]	00	✪Kigali*
St. Kitts & Nevis	[1]	011	✪Basseterre (869)**
St. Lucia	[1]	011	✪Castries (758)**

Key to International Access and Dialing Codes

✪ Capital city
†† More than one time zone in this country.
* City/area codes not required in this country.
** This city/area code used for entire country or territory.

Country	Country Code	International Access Codes	Capital City City/Area Code
St. Vincent & Grenadines	[1]	011	✪Kingstown (784)**
San Marino	[378]	00	✪San Marino*
São Tomé & Principe	[239]	00	✪São Tomé*
Saudi Arabia	[966]	00	✪Riyadh (1)
Senegal	[221]	00	✪Dakar*
Serbia & Montenegro	[381]	99	✪Belgrade (11)
Seychelles	[248]	00	✪Victoria
Sierra Leone	[232]	00	✪Freetown (22)
Singapore	[65]	001	✪Singapore*
Slovak Republic	[421]	00	✪Bratislava (2)
Slovenia	[386]	00	✪Ljubljana (1)
Solomon Islands	[677]	00	✪Honiara*
Somalia	[252]	00	✪Mogadishu (1)
South Africa	[27]	09	✪Cape Town (21) ✪Pretoria (12)
Spain	[34]	00	✪Madrid* (all nine-digit numbers; city codes are incorporated into local numbers)
Sri Lanka	[94]	00	✪Colombo (1)
Sudan	[249]	00	✪Khartoum (11)
Suriname	[597]	00	✪Paramaribo*
Swaziland	[268]	00	✪Mbabane* ✪Lobamba*
Sweden	[46]	00	✪Stockholm (8)
Switzerland	[41]	00	✪Bern (31)
Syria	[963]	00	✪Damascus (11)
Taiwan	[886]	002	✪Taipei (2)
Tajikistan	[992]	8*10	✪Dushanbe (37)
Tanzania	[255]	000	✪Dar es Salaam (22)
Thailand	[66]	001	✪Bangkok (2)
Togo	[228]	00	✪Lomé* (all seven-digit numbers)
Tonga	[676]	00	✪Nukualofa**
Trinidad & Tobago	[1]	011	✪Port-of-Spain (868)**
Tunisia	[216]	00	✪Tunis (1)

Country	Country Code	International Access Codes	Capital City City/Area Code
Turkey	[90]	00	✪Ankara (312)
Turkmenistan	[993]	8*10	✪Ashgabat (12)
Turks & Caicos	[1]	011	✪Grand Turk (649)**
Tuvalu	[688]	00	✪Funafuti*
Uganda	[256]	00	✪Kampala (41)
Ukraine	[380]	8*10	✪Kiev (44)
United Arab Emirates	[971]	00	✪Abu Dhabi (2) Dubai (4)
United Kingdom	[44]	00	✪London (20)
United States ††	[1]	011	✪Washington, DC (202)
Uruguay	[598]	00	✪Montevideo (2)
Uzbekistan	[998]	8*10	✪Tashkent (71)
Vanuatu	[678]	00	✪Port Vila*
Venezuela	[58]	00	✪Caracas (212)
Vietnam	[84]	00	✪Hanoi (4)
Virgin Islands (U.K.)	[1]	011	✪Road Town (284)**
Virgin Islands (U.S.)	[1]	011	✪Charlotte Amalie* (340)
Western Sahara	[212]	00	✪Laayoune (4)
Western Samoa	[685]		✪Apia*
Yemen	[967]	00	✪Sana'a (1)
Zambia	[260]	00	✪Lusaka (1)
Zimbabwe	[263]	00	✪Harare (4)

Key to International Access and Dialing Codes
✪ Capital city
†† More than one time zone in this country.
* City/area codes not required in this country.
** This city/area code used for entire country or territory.

Country Codes by Code

Code	Country	Code	Country
1	United States	64	New Zealand
1	Canada	65	Singapore
1 (264)	Anguilla	66	Thailand
1 (268)	Antigua & Barbuda	81	Japan
1 (242)	Bahamas	82	Korea, South
1 (246)	Barbados	84	Vietnam
1 (441)	Bermuda	86	China (PRC)
1 (345)	Cayman Islands	90	Turkey
1 (767)	Dominica	91	India
1 (809)	Dominican Republic	92	Pakistan
1 (473)	Grenada	93	Afghanistan
1 (671)	Guam	94	Sri Lanka
1 (876)	Jamaica	95	Myanmar (Burma)
1 (664)	Montserrat	98	Iran
1 (670)	N. Mariana Islands	212	Morocco
1 (787)	Puerto Rico	212	Western Sahara
1 (869)	St. Kitts & Nevis	213	Algeria
1 (758)	St. Lucia	216	Tunisia
1 (784)	St. Vincent & Grenadines	218	Libya
1 (868)	Trinidad & Tobago	220	Gambia, The
1 (649)	Turks & Caicos	221	Senegal
1 (284)	Virgin Islands (U.K.)	222	Mauritania
1 (340)	Virgin Islands (U.S.)	223	Mali
7	Chechnya	224	Guinea
7	Kazakhstan	225	Côte d'Ivoire
7	Russia	226	Burkina Faso
20	Egypt	227	Niger Republic
27	South Africa	228	Togo
30	Greece	229	Benin
31	Netherlands	230	Mauritius
32	Belgium	231	Liberia
33	France	232	Sierra Leone
34	Spain	233	Ghana
36	Hungary	234	Nigeria
39	Italy	235	Chad
40	Romania	236	Central African Rep.
41	Switzerland	237	Cameroon
43	Austria	238	Cape Verde
44	United Kingdom	239	São Tomé & Principe
45	Denmark	240	Equatorial Guinea
46	Sweden	241	Gabon
47	Norway	242	Congo, Rep of (Brazzaville)
48	Poland	243	Congo, Democratic Republic of (Kinshasa)
49	Germany		
51	Peru		
52	Mexico	244	Angola
53	Cuba	245	Guinea-Bissau
54	Argentina	246	Diego Garcia
55	Brazil	248	Seychelles
56	Chile	249	Sudan
57	Colombia	250	Rwanda
58	Venezuela	251	Ethiopia
60	Malaysia	252	Somalia
61	Australia	253	Djibouti
62	Indonesia	254	Kenya
63	Philippines	255	Tanzania

256	Uganda	595	Paraguay
257	Burundi	596	Martinique
258	Mozambique	597	Suriname
260	Zambia	598	Uruguay
261	Madagascar	599	Netherlands Antilles
262	Reunion	673	Brunei
263	Zimbabwe	675	Papua New Guinea
264	Namibia	676	Tonga
265	Malawi	677	Solomon Islands
266	Lesotho	678	Vanuatu
267	Botswana	679	Fiji
268	Swaziland	680	Palau
269	Comoros	682	Cook Islands
297	Aruba	684	American Samoa
298	Faeroe Islands	685	Western Samoa
299	Greenland	687	New Caledonia
350	Gibraltar	688	Tuvalu
351	Portugal	689	French Polynesia
352	Luxembourg	692	Marshall Islands
353	Ireland	808	Midway Islands
354	Iceland	850	Korea, North
355	Albania	852	Hong Kong
356	Malta	853	Macau
357	Cyprus	855	Cambodia
358	Finland	856	Laos
359	Bulgaria	880	Bangladesh
370	Lithuania	886	Taiwan
371	Latvia	960	Maldives
372	Estonia	961	Lebanon
373	Moldova	962	Jordan
374	Armenia	963	Syria
375	Belarus	964	Iraq
376	Andorra	965	Kuwait
377	Monaco	966	Saudi Arabia
378	San Marino	967	Yemen
380	Ukraine	968	Oman
381	Kosovo	971	United Arab Emirates
381	Serbia & Montenegro	972	Israel
385	Croatia	973	Bahrain
386	Slovenia	974	Qatar
387	Bosnia & Herzegovina	975	Bhutan
389	Macedonia	976	Mongolia
420	Czech Republic	977	Nepal
421	Slovak Republic	992	Tajikistan
423	Liechtenstein	993	Turkmenistan
500	Falkland Islands	994	Azerbaijan
501	Belize	995	Georgia
502	Guatemala	996	Kyrgyzstan
503	El Salvador	998	Uzbekistan
504	Honduras		
505	Nicaragua		
506	Costa Rica		
507	Panama		
509	Haiti		
590	French Antilles		
590	Guadeloupe		
591	Bolivia		
592	Guyana		
593	Ecuador		
594	French Guiana		

Cellular (Mobile) Phones

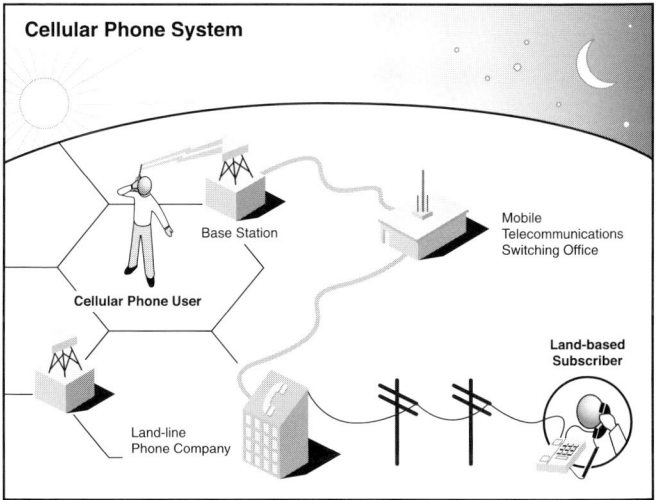

Cellular Phone System

Base Station

Cellular Phone User

Mobile Telecommunications Switching Office

Land-based Subscriber

Land-line Phone Company

What Is a Cellular Telephone?

Cellular (mobile) telephones are radio devices that send and receive communications signals (voice and data) via networks of earth-bound telecommunications stations distributed over geographical regions called cells.

How Do They Work?

The Handset When you make a call the cell phone (called a handset) converts your voice, fax or data into a radio signal that is transmitted to a base station.

The Cell Cellular service systems are comprised of a set of geographic cells that each contain a base station. Each cell's service boundaries are defined by the strength of the station's radio transceiver and the technology used, as well as by the local terrain.

The Base Station Base stations receive, manage and send signals from mobile radio telephone units. The signals are routed to cellular telephone switches, which in turn route them to a main switching office. The main office routes calls between mobile units, to land-based telephones, and from land-based telephones to mobile units. As a mobile unit moves from cell to cell, the main switching office switches the call between cells.

When you turn on your phone it sends a signal to the closest base station where the system checks your subscriber number and data. The phone checks for a signal from the base station by scanning the radio frequencies and testing reception. The system switches a call to the closest cell based on signal strength.

Analog Phones A coder inside the phone converts sound to analog electrical signals, representing speech as a continuous electromagnetic wave. These waves are transmitted and at the point of reception reconverted to sound. **Note:** Analog technology is being abandoned and analog networks are no longer in construction, although there are vast areas still covered by analog service.

Digital Phones Sound is converted into a digital signal inside the phone. The signal is then sent and reconverted to sound or data at the point of reception. All new and future networks use digital technology, which has undergone and will continue undergoing extraordinary enhancements.

How Well Do They Work?

Cellular telephones operate only within range of a base station in a given cell. As a caller moves from cell to cell, it is possible to lose the connection if the call cannot be switched to a new station. Also, the caller may experience a short interruption or radio static during the switching process. There have been impressive gains in quality from digital technologies, with more coming in future years.

Radio transmissions are difficult in mountainous areas. Vast areas of the world remain without cellular coverage, although major metropolitan areas that are covered by new, digital networks have excellent coverage.

Other Uses of Cell Phones

Text-messaging Sending and receiving short, text messages has become a popular function of cell phones, especially outside the U.S. See "SMS & MMS" on page 47.

What's HOT Mid 2005

Here are the latest, greatest multi-band GSM/CDMA cell phones from major manufacturers as of March 2005.

Motorola RAZR V3

Quad-band GSM, Bluetooth, MP3 ringtones, speakerphone, SMS, EMS, MMS, WAP browser, GPRS, MPEG4 video playback, external display, built-in VGA camera with 640 x 480 pixel resolution and 4x digital zoom; mini USB connection; adapter; completely flat illuminated keypad, anodized aluminum case. 0.54 in. thick; 3.35 oz.

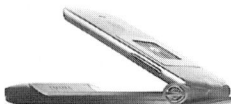

LG VX8000

1.9 Ghz CDMA PCS, 800 MHz CDMA (all digital); 1.3 megapixel CCD camera with flash and 1280 x 960 pixel resolution, 4x digital zoom, MP3 player, speakerphone, video/still image capture, support for high-speed EVDO 3G networks (make sure Verizon's EVDO network is available in your area), V CAST service to stream video clips (up to 3 minutes); 3.9 oz.

Sony Ericsson Z800i

Tri-band GSM/GPRS (900/1800/ 1900), Bluetooth, WCDMA 2100 MHz, Infrared, USB, SMS, MMS, e-mail, video call functionality, 1.3 megapixel camera, multi-format music player, UMTS radio, 4.5 oz.

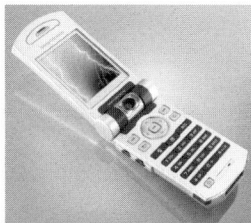

Internet and E-mail The latest cell phone technology allows users to serf the Internet and send and receive e-mail messages with multi-media attachments.

Photos Many cell phones now enable the user to send and receive low-resolution photos taken with either a built-in camera or small attachment.

Audio/Video Third-generation technology allows users to listen to audio and view full-color video.

Keeping Time Cell phones can be used by travelers as alarm clocks. Cell phones display time, synchronizing automatically with the time zone of the cellular network that they are connected to.

Where Can You Purchase One?

Cell phones can be purchased from retail telephone, electronics, and appliance stores as well as on the Internet from manufacturers or third-party suppliers. Airtime plans are usually bought at the same time as the phone. Comparison shop at sites such as www.amazon.com.

How Much Do They Cost?

Hardware Cellular telephones cost from US$50 to US$500. Many companies offer special promotions for free or nominal cost phones in exchange for a one or two year airtime contract. Accessories include spare batteries, headphones and chargers. **Warning!** Be very wary of GSM cell phone promotions, because most give the user a phone which is locked and cannot be used with any other provider's SIM card. Request and obtain the unlock code, keeping in mind that the unlock code cannot be entered until you insert a different provider's SIM card.

Service Plan/Airtime Monthly plans include "anytime" and "off-peak" minutes. In the U.S. many plans are available for US$20 to US$50 per month. Additional airtime is charged on a per minute basis. Extra charges are added for calls made from outside the provider's service area.

GSM cell phone users also have the option of purchasing "prepaid SIM (Subscriber Identification Module) cards" and using these to handle on-air charges. See "Cell Phone SIM Cards" on page 32 for more information on this popular option.

In addition to purchasing a local SIM card, purchase a prepaid calling card. The minutes you use under the calling card are much cheaper; for example, calling from the UK to the U.S. is as low as US$0.08 per minute.

Cellular Service Types

There are a number of competing cellular technologies in the world. The current leader is GSM (Global System for Mobile Communications). See the tables on the following pages. One clear trend is that the older analog networks are rapidly being abandoned in favor of digital systems.

Will My Phone Work Overseas?

This depends upon whether (1) Your cell phone covers the frequency range used overseas. (2) your cell phone type (e.g., GSM, CDMA, etc.) exists in the country of intent. (3) If (1) and (2) are satisfied, and if you want to use your home service provider and phone number, your service provider must have a reciprocal agreement. (4) If you use a tri-band GSM phone and there is GSM service where you go, you may elect to purchase a local SIM card for your phone.

If you are interested in using your home provider while overseas, contact your provider prior to departure. Also, refer to "U.S. Cellular Provider Table" on page 41. Also, see "Renting a Cell Phone While Traveling" on page 34.

Examples: You are a Canadian traveler to Europe. If you use a GSM 1900 phone in Canada, you have to make sure that it also works with the 900 mhz signals used in most of Europe; otherwise, rent a GSM tri-band phone. The same traveler, this time to Japan, must rent or purchase a CDMA cell phone for use in Japan, where most networks use CDMA, not GSM.

Cellular Service Systems Worldwide

Region	Operating System
Africa	**AMPS, CDMA, GSM, TACS** GSM 900 networks predominate by far. A few older British analog TACS networks exist in some of the British Commonwealth countries (Kenya, for example), though GSM 900 likewise predominates in those countries. A few CDMA networks exist (in Zambia, for example), but most of Africa will continue piggy-backing the European GSM technology.
Asia	**AMPS, CDMA, GSM, TACS** GSM, especially on the 900 mhz frequency, is the predominant technology outside of Japan and South Korea. Indonesia has various technologies, including older AMPS and newer CDMA and GSM networks. Presumably, current GSM networks will migrate to European 3G UMTS standards and technology.
Australia	**CDMA, GSM** GSM 900 is the predominant technology, with newer CDMA networks coming online. A 3G network is being built, which combines GSM and CDMA technologies with improved data-transmission capacities.
Canada	**AMPS, CDMA, GSM, TDMA** Canada's history of cellular networks parallels North America as a whole. The original analog AMPS technology has largely been replaced, although modified AMPS networks still support use of digital TDMA (the original digital enhancement of AMPS) phones; the TDMA networks are being replaced by GSM. As in US, all GSM networks are on the 1900 mhz frequency. CDMA networks are likewise 1900 mhz. Upcoming 3G systems employ CDMA2000 technology.
Central America	**AMPS, CDMA, GSM, TDMA** AMPS and TDMA (the original digital enhancement of AMPS) are still predominant. CDMA and GSM are gradually replacing the older AMPS-based networks. Both the European 900/1800 and the North American 1900 frequencies are appearing.
China	**CDMA, GSM** China currently has GSM 900/1800 and CDMA 900/1800 networks exclusively.
Europe	**AMPS, GSM, NMT** GSM 900 MHz is the predominant system currently used (now adding 1800 MHz). Analog AMPS and NMT networks are obsolete. UMTS (using enhanced 3G GSM and CDMA technologies) will be the standard by the middle of the decade.
India	**GSM** India currently uses GSM 900 networks exclusively.

Region	Operating System
Japan	**CDMA, PDC** PDC, based on TDMA (the original digital enhancement of AMPS), was predominant in Japan, but newer-generation CDMA2000 and W-CDMA networks now predominate. Japan is fast moving to 3G networks, which combine GSM and CDMA technologies with improved data-transmission capacities.
Mexico	**AMPS, CDMA, GSM, TDMA** Mexico's history of cellular networks parallels North America as a whole. The original analog AMPS technology has largely been replaced, although modified AMPS networks still support use of digital TDMA (the original digital enhancement of AMPS) phones; TDMA networks are being replaced by GSM. As in US, all GSM networks are on the 1900 mhz frequency. CDMA networks are likewise 1900 mhz.
Middle East	**AMPS, CDMA, GSM, NMT, TACS, TDMA** GSM 900 networks predominate by far. Analog AMPS networks are obsolete. In addition to GSM 900/1800, Israel has several CDMA networks and a TDMA network.
Russia	**CDMA, GSM** GSM 900/1800 networks predominate. Several CDMA networks exist.
Scandinavia	**GSM, NMT** NMT was predominant in 80's and 90's but has largely been replaced by GSM 900 and 1800 networks.
South America	**AMPS, CDMA, GSM, TDMA** AMPS and TDMA (the original digital enhancement of AMPS) are still predominant, especially in poorer countries. GSM networks on the European 1800 mhz frequency are being integrated into Brazil, with GSM 1900 mhz (North American GSM standard) being integrated in Argentina and Chile. CDMA networks are also being built
United States	**CDMA, GSM** Not only are the original analog AMPS networks obsoleted, but the original digital TDMA (the original digital enhancement of AMPS) networks have been replaced by country-wide GSM networks operating on the 1900 mhz frequency. CDMA 1900 networks are the most numerous. Upcoming 3G systems will employ CDMA2000 technology.

Digital Cellular Technologies Worldwide

Service	Definition	Range
CDMA	**Code Division Multiple Access** Originally developed by Qualcomm ("CDMAOne"). Considered the best single cellular technology for efficiency and quality of sound, but requires a higher investment in infrastructure. Used predominantly in USA, Canada, South Korea, Japan, and China. CDMA splits radio signals over different channels, which are re-assembled into a single channel on the receiving end.	800 mhz 1900 mhz
CDMA 2000	**Code Division Multiple Access 2000** Like W-CDMA, CDMA2000 is a 3G standard that uses a broader spectrum than CDMAOne. CDMA2000 is similar, but not the same as W-CDMA, the other CDMA-based 3G standard. CDMA2000 is considered slightly more advanced than the competing W-CDMA technology.	1900 mhz 2100 mhz
CDPD	**Cellular Digital Packet Data** Standard developed by a consortium of US companies in the late 1990s. Uses channel hopping to send data in short bursts during idle times in cellular channels. Piggybacks and enhances AMPS analog systems. Permits TDMA phones to transmit data in addition to voice.	800 mhz 900 mhz
CTS	**Cordless Telephone System** European-developed add-on to GSM systems. The purpose of GSM CTS is to exploit the sound quality and inexpensive use of land-line systems while getting the services and mobility of GSM.	900 mhz 1800 mhz
D-AMPS	**Digital AMPS (Advanced Mobile Phone Service)** Uses TDMA. technology. Although currently being replaced by CDMA and GSM 1900 networks, D-AMPS has been widely used in the Americas because the original analog AMPS networks permit the use of D-AMPS cell phones.	800 mhz
DCS 1800	**Digital Cordless Standard** Original reference to GSM 1800	1800 mhz
EDGE	**Enhanced Data Rates for GSM Evolution** Technology enabling 3G (Third Generation) data rates. Builds upon GPRS. Component of the migration of GSM and TDMA to 3G UMTS. Employs 16 Quadrature Amplitude Modulation (16QAM), rather than GSM Gaussian modulation shift keying (GMSK). Requires additional base stations and infrastructure for established GSM networks migrating to EDGE. Enables services such as multimedia e-mailing and video conferencing from wireless terminals.	850 mhz 900 mhz 1800 mhz 1900 mhz
E-Netz	Alternate German reference to GSM 1800 networks	1800 mhz

Service	Definition	Range
E-TDMA	**Extended Time Division Multiple Access** Enhancement of TDMA that splits the fixed number of radio frequencies into more time slots, allowing more simultaneous phone calls.	900 mhz
GMSS	**Geostationary Mobile Satellite Standard** This standard was developed from cellular GSM	1800 mhz
GPRS	**General Packet Radio Service** Important enhancement to GSM technology that increases data transmission by the use of data packets. GPRS has enabled GSM cellular phones to make calls and transmit data at the same time. When introduced in the late 1990s, GPRS increased data transmission speeds by a factor of three in comparison with landline telephone networks, and by a factor of 10 in comparison with the existing Circuit Switched Data services on GSM cellular networks.	900 mhz 1800 mhz 1900 mhz
GSM	**Global System for Mobile Communications (formerly "Group Speciale Mobile")** European standard that is the most widely used cellular technology around the world. Uses a modified and very efficient TDMA technology at its base. Earlier 900 mhz systems are still common inside and outside of Europe. 1800 mhz systems are found in Europe and with increasing frequency elsewhere, although Europe is moving rapidly to 3G (Third Generation) networks, which will combine enhanced GSM and enhanced CDMA technologies. North American GSM networks are all on the 1900 mhz frequency.	900 mhz 1800 mhz 1900 mhz
iDEN	**Integrated Dispatch Enhanced Network** Enhanced TDMA technology by Motorola. Used in North America and parts of South America and Asia.	800 mhz 900 mhz
N-CDMA	**Narrowband CDMA (Code Division Multiple Access)** The original CDMA (IS-95) developed by Qualcomm. Also referred to as CDMAOne	800 mhz
PCS	**Personal Communications System** **Digital Communications System** Name for various wireless technologies and services (predominantly in North America) operating in the 2 GHz range. The US Federal Communications Commission has allocated 140 mhz of space between 1850 mhz and 1990 mhz for broadband PCS. The GSM and CDMA 1900 networks in North America are also referred to as PCS or DCS 1900 systems.	1850 to 1990 mhz
PDC	**Personal Digital Cellular** A 2G (Second Generation) Japanese technology that uses a variation of TDMA different from the TDMA used in the European-standard GSM networks. First appeared in 1991, and only used in Japan. Will be obsoleted by 3G (Third Generation) systems and standards.	800 mhz 1500 mhz

Service	Definition	Range
SDMA	**Spacial Division Multiple Access** Special technology that increases capacity of every type of current or future cellular system. Uses antenna arrays at base stations. SDMA directs signals towards the mobile unit that is the intended target, rather than dispersing them throughout a cell's area as traditional systems do.	all current frequencies
TDMA	**Time Division Multiple Access** An early digital technology, enhancements of which are still widely used, especially in GSM systems. Squeezes multiple conversations into one cellular channel. Original TDMA is still found in Central and Latin America, since TDMA phones can function using original analog AMPS cellular networks.	900 mhz
TETRA	**Terrestrial Trunked Radio** Standard created by the European Telecommunications Standardisation Institute (ETSI)	any frequency below 1000 mhz
UMTS	**Universal Mobile Telecommunications Standard**. A 3G (Third Generation) European standard based on W-CDMA. Uses the 2Ghz band. Intended as a world-wide standard for European GSM networks of the future.	All European GSM frequencies
W-CDMA	**Wideband CDMA (Code Division Multiple Access)** Not to be confused with the similar CDMA2000. W-CDMA is a considerable enhancement of the original Qualcomm CDMA. Integral part of 3G (Third Generation) UMTS standard. Has 10 times the computational capacity of 2G (Second Generation) technologies.	1900 mhz 2100 mhz

Notes

Analog Cellular Technologies Worldwide

ANALOG CELLULAR TECHNOLOGIES WORLDWIDE		
Service	**Definition**	**Range**
AMPS	**Advanced Mobile Phone Service** Developed in U.S.A. by Bell Labs in the 1970's, with first commercial network in 1983. Formerly very prevalent in the Americas, but rapidly being phased out.	800 mhz
C-450	Older South African technology from 1980's, operating under Vodacom. Obsoleted by GSM and other digital technologies	450 mhz
C-Netz	Early technology employed in Germany and Austria. Obsoleted by GSM.	450 mhz
Comvik	Early (1981) technology built and used in Sweden for the Comvik network. Replaced in Sweden by GSM.	450 mhz
FDMA	**Frequency Division Multiple Access** The basis of the original AMPS analog systems, considered an inefficient technology that has largely been replaced by newer digital systems. Each call is on a different, reserved frequency.	800 mhz
N-AMPS	**Narrowband Advance Mobile Phone System.** Development by Motorola of AMPS employing 10khz channels (the 30-khz AMPS bandwidth is divided into three 10-khz channels). Increased AMPS capacity three-fold.	800 mhz
NMT	**Nordic Mobile Telephony** Original Scandinavian system designed to work in uneven, hilly terrain. Was installed in many countries around the world; obsoleted by digital GSM technology.	450 mhz, 900 mhz
NTT	**Nippon Telegraph and Telephone.** Original analog technology used in Japan. Obsoleted by digital CDMA technology.	450 mhz 800 mhz
RC2000	**RadioCom 2000** French system made operative in 1985. Obsoleted by digital GSM technology	n/a
TACS/ ETACS	**Total Access Communications System / Extended TACS** AMPS with a few minor changes, operating in the 900 mhz frequency band. Developed by Motorola, though primarily U.K.-based. Obsoleted by digital GSM technology.	900 mhz

GSM "World" Phones

The Issue

As the mobile phone industry developed in the 1970s, 80s and 90s, telecommunications firms worldwide raced to develop what they thought would be the dominant technology. Some early systems didn't work very well and were replaced by better and more modern systems. There were and still are, however, many cellular systems in the world.

As a result, international travelers face the issue of incompatibility of cell phone hardware (the phone itself) and cellular technology (service systems) operating in different countries and regions of the world.

In the past few years, there has been a "shake out" in both the provider industry as well as in underlying technologies. Financially weak cellular providers have been replaced by stronger businesses. Also, older cellular technologies have been replaced by modern systems, and more specifically, the (currently) dominant technology of GSM.

What Is GSM?

GSM is an acronym for Global System for Mobile Communications. GSM was originally a European standard established in the mid 1980s. It is now the most widely used digital cellular technology in the world with 60 percent of the market.

Technically, GSM uses a modified and very efficient combination of FDMA (Frequency Division Multiple Access) and TDMA (Time Division Multiple Access) technology at its base. Earlier 900 mhz systems are still common inside and outside of Europe, but 1800 mhz systems are dominant in Europe and with increasing frequency elsewhere in the world. Europe is moving rapidly to 3G (Third Generation) networks, which combine enhanced GSM and enhanced CDMA (Code Division Multiple Access) technologies. North American GSM networks are all on the 1900 mhz frequency.

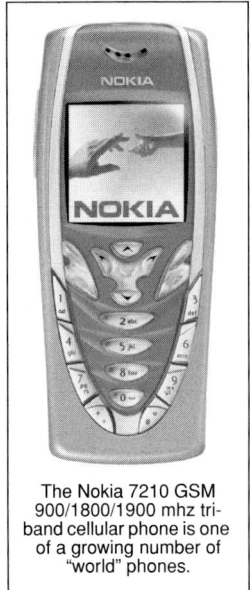

The Nokia 7210 GSM 900/1800/1900 mhz tri-band cellular phone is one of a growing number of "world" phones.

What Does a GSM "World" Phone Do for the User?

Unlike other cell phones where the identity of the phone is part of the instrument itself , the identity (and phone number) of a GSM phone is in a user-removable smart card known as the SIM card (subscriber identification module). This enables a GSM cell phone user to carry out several options. It allows the user to:

- carry his SIM card and insert it into an in-country purchased new phone, (or)
- purchase a new SIM card for his existing phone, (or)
- use his phone and SIM card "as is", (or)
  buy both in the country being visited.

A GSM cell phone allows a user with a service provider with extensive roaming agreements to travel the world without changing phones. Other phone systems may allow the user this capability, but not as well as GSM. This is especially the case in Europe, the USA, Middle East, Russia, India and China.

What Do I Need?

To get the international travel benefits of GSM you will need two things:

1. **A GSM phone**. Many users who travel purchase tri-band phones which allow them to utilize any version of GSM service as they travel. Be aware that different GSM technologies use different frequencies. We recommend a GSM tri-band 900/1800/1900 mhz phone. Ask your provider what phone is best based on the countries you plan to visit.

   Note: Be sure to get a GSM phone that does not have a SIM card "lock." This will prevent you from purchasing local airtime SIM cards as you travel. See "Cell Phone SIM Cards" on page 32 for details.

2. **A service provider utilizing GSM roaming agreements**. Ask potential providers for specific information about coverage and roaming agreements in the countries in which you plan to travel.

   Note: Rather than using your home country service provider and international roaming, you may be better off getting a GSM phone with an unlocked SIM card and purchase local airtime SIM cards as you travel. See "Cell Phone SIM Cards" on page 32 for more details.

Notes

Cell Phone SIM Cards

What Are SIM Cards?

A SIM (Subscriber Identification Module) card is a detachable device that, when inserted into a GSM cell phone, activates the phone with account information. Essentially, a SIM card is the key enabling a subscriber to access and use GSM networks around the world.

Although GSM cards come in what seem like two different sizes, credit card size and a small 1 sq. in. size, the credit card-sized card can be readily converted into the small one by popping out the small-size version from the rest of the useless plastic.

How Do SIM Cards Work?

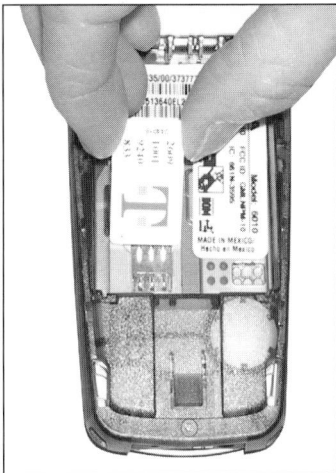

Each SIM card has a significant (64K) amount of memory embedded on the computer chip that stores information relating to the subscriber: phone number, phone settings, personal security keys as well as account information such as rate plans and service features. The SIM card identifies the subscriber to the cellular network and assures proper billing of the user's account.

Once inserted into a GSM cell phone, that phone becomes the subscriber's phone. The phone's number automatically changes to the number on the SIM card.

SIM Security Features

Most SIM cards use a Personal Identification Number (PIN) that is stored on the card by the subscriber. After the SIM card is inserted into a GSM handset, the PIN number must be entered to activate the phone. This is similar to ATM-required PINs. Since many users dispense with the option of having to enter a PIN number every time the phone is turned on, a lot of potentially sensitive contact information can be compromised if the SIM card or phone is stolen.

Prepaid SIM Cards

Prepaid cellular services have become extremely popular. A prepaid SIM card allows a traveler to initiate calls for a specific, cumulative number of minutes. It lets the subscriber pay only for time spent making calls and avoids minimum monthly fees and other trappings of traditional accounts. Overusage is also minimized, because the user is reminded of the level of usage every time a new card is purchased (very much like filling a gas tank).

Often, but not always, prepaid SIM cards are limited to the country in which they are purchased. Also, purchase may require identification. Finally, some prepaid SIM cards now include a certain amount of free calling time.

Recharging a Prepaid SIM Card

SIM cards typically expire 12 months or less after purchase. If you use up your allotted minutes it's easy to "recharge" the card by purchasing "airtime vouchers" at a cell phone shop or vending machine. The majority of European users recharge their SIM cards. To recharge, dial the toll-free number on the voucher and follow the automated instructions. Once you are done, your card is automatically recharged with the minutes purchased.

Prepaid Card vs. User's Own SIM Card

Should a traveler abroad bring along their own SIM card or purchase prepaid cards upon arrival? Answers depend on the traveler's needs.

The main advantage to using one's own SIM card is that any call to the traveler's home cell phone number will automatically connect to the traveler's overseas cell phone. No one needs to be notified of the traveler's whereabouts or of a special new phone number.

The drawback to using one's own SIM card abroad is a much higher cost even for local in-country calls and that an international roaming agreement is required with the traveler's home provider. These agreements are usually more costly than calls made using prepaid cards provided by the overseas networks. Also, under roaming agreements, incoming calls are chargeable to the traveler, but free, or at very low cost, when using a prepaid card.

SIM Card
Actual size.

Some savvy travelers recommend a compromise: bring along the home SIM card and keep it inserted in the phone most of the time. When outgoing calls are made, remove the home SIM card and insert the prepaid card.

Renting vs. Purchasing

If the traveler's own GSM cell phone is incompatible with network(s) in the foreign countries, the user must either purchase or rent a compatible phone. Experienced travelers recommend the purchase, not rental, of a compatible phone. In some instances, renting a phone for one week costs as much as the purchase of that type of phone. See "Renting a Cell Phone While Traveling" on page 34 for more information.

SIM Backups

A recent, user-friendly development is the ability to "back up" information from a SIM card. The backup is on a small, digital device that is easily transportable. If the user's phone is lost or stolen, the backup can be used to write the information to a new SIM card.

Checklist for Using Prepaid SIM Cards

1. **Obtain Compatible GSM Phone.** Purchase, rent or borrow a phone that is compatible with the GSM network(s) in the countries you plan to visit.
2. **Purchase SIM Card.** You can purchase these online from many cellular carriers or wait until you arrive and purchase cards at a convenience stores. Make sure the card you purchase will fit into your phone. If you purchase an in-country prepaid SIM card, you may not be able to activate it until after your reach that country.
3. **Insert SIM Card.** Make a note of the phone number associated with the card. The number is usually displayed on the packaging, not on the SIM card itself.
4. **Enter Appropriate Codes.** SIM cards come with brief instructions on what to do when inserting and activating your GSM cell phone.
5. **Make Calls**.
6. **Purchase "Airtime" Vouchers.** Purchase these vouchers at phone and convenience stores. Dial the number found on the voucher and follow the automated instructions.

Notes

Renting a Cell Phone
While Traveling

Renting a GSM "world" cell phone for international travel has become very popular and a big business, especially via online Internet rental stores. There are issues, however, that the smart traveler needs to address, including pitfalls that can cost the uninformed traveler lots of money. Here are the particulars.

Will Your Existing Phone Work?

Before you consider renting a phone, check to see if your existing phone and service will work in the countries you are visiting. Check with your provider regarding the countries you are visiting and the cell phone you have. You may only need an international roaming agreement. Beware that these roaming charges may be very high. In fact, if you use your home-purchased GSM service provider and SIM card abroad, you may pay the same high flat rate per minute to call across the street as you would pay to call across the world.

Home Provider Rentals

As a general rule, do not rent a cell phone from your home provider. The pitfalls of these rentals are numerous: 1) rental fees are higher than third-party online stores or convenience stores in the foreign countries, 2) Roaming agreements between the home provider and providers in the foreign countries always involve higher per-minute charges. 3) You are likely to be charged for incoming calls, even those that are not picked up. 4) Many roaming agreements require a minimum usage charge per day, so you end up paying a daily fee even when you make no calls. 5) These phones are often "SIM card locked" so you must use their proprietary—and more expensive—SIM cards.

Rent or Purchase?

If you travel on a regular basis, we recommend that you purchase and not rent a cell phone. Most rented cell phones bill by the airtime minute, which cannot tell if a call was answered, busy, or not, because renting companies don't have access to that information from foreign cellular providers. Renters end up paying huge amounts for "busy" and "no answer" calls rounded up to the nearest minute. For travelers to GSM countries, we recommend that you purchase a tri-band GSM 900/1800/1900 phone. This will work throughout Europe and all non-European GSM countries, including GSM 1900 networks in North America.

It pays to shop online for your phone. If you purchase a phone overseas, you may get hit—as you would in most European countries—by a VAT (value-added-tax) that can substantially increase the retail sales price. Some countries may refund the VAT, if you supply the requisite paperwork and receipts, when you leave that country.

Also, make sure the supplier re-programs the phone's menus in your language. For example, a U.S. traveler who rents or buys a phone in Germany may have to read the menus in German. Many suppliers will make this change without being asked, but check the displayed language before heading out and using the phone.

SIM Cards

A SIM (Subscriber Identification Module) card is a detachable device that, when inserted into a GSM cell phone, activates the phone with personal and account information. The SIM card is the key that enables a GSM subscriber to access and use GSM networks around the world.

Our second recommendation is that you purchase pre-paid SIM cards for the countries you will be visiting. As a result, make sure that the GSM phone you purchase is not SIM card locked. If it is SIM-card-locked, only SIM cards approved by the provider (in agreement with the foreign providers, if any) will work. Per-minute charges using such SIM cards are substantially higher than charges using prepaid SIM cards purchased locally. Also, some prepaid SIM cards may allow you to receive phone calls free of charge while abroad (some prepaid SIM cards allow for a limited amount of airtime minutes, whether incoming or outgoing). Prepaid SIM cards, in most circumstances, are such a good idea that Europeans make the majority of cell phone calls using such prepaid cards.

See "Cell Phone SIM Cards" on page 32 for details.

Online Rental Providers

If you want to rent or purchase a phone for use overseas, do so from a reputable online service, not from your local cellular provider. Specify that the phone be "unlocked". You can also purchase prepaid SIM cards from such online services, although it is very easy to purchase prepaid cards in the destination countries.

Example of Different Rates: A U.S. traveler to Europe using a U.S. provider with a European roaming agreement will pay US$0.99 per minute for outgoing calls and, depending on the rental agreement, the same for incoming calls.

The same traveler who obtains an unlocked GSM phone and purchases prepaid SIM cards pays US$0.45 per minute for calls to the U.S., and substantially less for calls within Europe itself.

When to Use Roaming Agreements

When you use roaming agreements, the cell phone you use overseas carries your "home" phone number, so that anyone calling this number will automatically ring you overseas, regardless of whether the caller knows your whereabouts or not. A prepaid SIM card, on the other hand, involves a "new" number. Thus, there is one scenario in which it makes sense to use roaming agreements: you are not really interested in making calls while overseas, but want to receive calls, in case of emergency, on your home cell phone number. An alternative strategy is to purchase prepaid SIM cards online in advance, rather than waiting until you are abroad. You then have time to inform necessary parties of the cell phone number(s) where you can be reached overseas.

Insurance on Rental Phones

Cell phone theft has become a real issue in many parts of the world. Without insurance, the rental supplier will charge you the maximum retail price for the lost or stolen phone. Insurance premiums vary; as an example, a major online rental agency charges US$2 per day per phone.

Security Issue: Your stolen GSM phone contains a record of outgoing and incoming phone calls, plus any number you have personally entered into that phone's memory.

Cell Phone Travel Tips

1. **Get the Right Cell Phone** Purchase, rent, borrow, or bring along a cell phone that is compatible with the network(s) that you will be using while abroad.

2. **SIM Cards** If you are traveling to GSM countries, consider purchasing a prepaid SIM card. (These can be purchased after you arrive in most convenience stores and kiosks; also available online from Internet stores.)

3. **Avoid Roaming Agreements** These agreements between your home provider and providers overseas will cost you significantly more for the use of a cell phone abroad. Try prepaid SIM cards instead. See "Renting a Cell Phone While Traveling" on page 34.

4. **Prepaid Calling Cards** Consider purchasing a prepaid calling card for use overseas. In conjunction with prepaid SIM cards, this option guarantees the lowest per-minute charges in most countries.

5. **Battery Charger** Bring the battery charger and make sure it works with local electrical outlets. While most cell phone chargers handle all usual voltage (100 to 240v), some can only handle the voltage of the country in which they were purchased. Check the wording on the back of the charger and get a transformer if needed; do NOT use a "converter" (identifiable from its power setting range of 50 to 1600w).

6. **Adapters and Plugs** Be sure to bring the appropriate adapters and plugs for your destination countries.

7. **Car Adapter** If you will be renting a car overseas, consider bringing a car cigarette lighter power adapter to charge your phone in the car.

8. **Laptop Computer Connector** If you will be bringing along or using a laptop computer, bring a connector to connect the cell phone to the laptop. Depending on the cell phone you have, you may or may not have to also establish a service agreement with the cellular service provider abroad for laptop data service. If the cell phone is used merely to connect to your ISP, then all you will need is the software "drivers" with the adapter. In most cases, however, you will need to establish (and pay for) a full digital data service "upgrade".

9. **Headset** It's always a good idea to bring along a headset, especially for driving in countries with laws requiring hands-free cell phone use while driving.

10. **Cell Phone Hands-Free Car Kit** For the best possible reception from mobile phones, especially in rural areas, carriers recommend the use of a mobile phone hands-free car kit with an external vehicle-mounted high gain antenna. Be sure to get the highest gain CDMA and GSM antenna available (the gain of the antenna indicates the antenna's ability to send and receive weak signals). Even so, do not place much faith in external antennas for cell phones because (1) at the typical 1800/1900 MHz band, the car's windows provide ample view of the outside world to your cell phone. (2) the signal attenuation in the coaxial cable of the external antenna at 1800/1900 MHz may offset any benefit from having an external antenna in the first place.

Cell Phone Travel Kit

The Kit

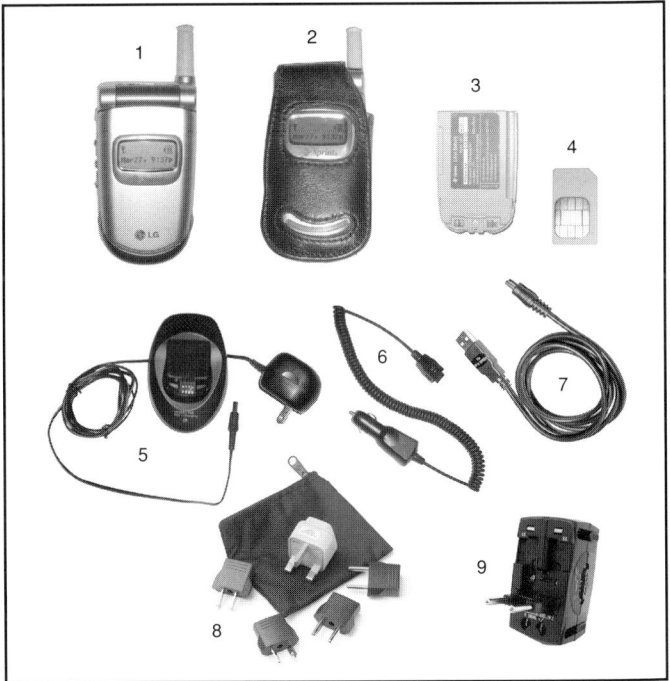

The Checklist

1. Cell Phone
2. Cell Phone Case
3. Extra Cell Phone Battery (1 - 2 recommended)
4. SIM Card (for GSM phones only)
5. Battery Charger
6. Cell Phone Car Adapter
7. Cell Phone Laptop Connector
8. Electric Plug Adapter Kit
9. Electric Power Converter
10. Connection to PDA (not shown)
11. Headset (not shown)

Notes

Cell Phone Rental Services

There are thousands of firms worldwide that rent cell phones. This list can serve as a start point.

Note: Most rented cell phones bill by the airtime minute, which cannot tell if a call was answered, busy, or not, because rental companies cannot access that information from foreign cellular providers. Renters need to know that they may end up paying huge amounts for "busy" and "no answer" calls, rounded up to the nearest minute.

United States

Dollar-Rent-A-Phone
Tel: [1] (212) 734-6344
Tel: [1] (800) 964-2468 (toll-free US/Canada)
Web: www.dollar-rent-a-phone.com
E-mail: mail@dollar-rent-a-phone.com

Europe-USA Cellular Phones Rentals
150 E. 69th St.
New York, NY 10021
USA
USA Phone Rental
Tel: [1] (800) 964-2468 (toll-free)
Tel: [1] (212) 734-6344

IMC Worldcell
International Mobile Communications, Inc.
801 Roeder Road, Suite 800
Silver Spring, MD 20910
USA
Tel: [1] (888) 967-5323
E-mail: orders@worldcell.com
Web: www.worldcell.com

International Phone Rental
Tel: [1] (888) 802-9518 (toll-free)
Tel: [1] (917) 774-3486
Web: www.europe-usa-mobile-cellular-phone-rental.com
E-mail: mail@europe-usa-mobile-cellular-phone-rental.com

InTouch Global
4100 B Westfax Dr.
Chantilly, VA 20151
USA
Tel: [1] (703) 222-7161
Tel: [1] (800) 872-7626 (toll-free in USA)
Web: www.intouchglobal.com

Planet Omni
Specializing in Global GSM Communication
1480 Wharton Way
Concord, CA 94521
USA
Tel: [1] (800) 858-4289 (toll-free US/Canada)
Tel: [1] (925) 246-7103
Fax: [1] (925) 686-9968
Web: www.planetomni.com
E-mail: info@quantumstar.com

Roberts Rent-A-Phone
150 E. 69th St.
New York, NY 10021
USA
Tel: [1] (800) 964-2468 (toll-free in USA)
Tel: (888) 802-9518 (toll-free in Europe)
Web: www.roberts-rent-a-phone.com

TripTel
1525 Van Ness Ave.
San Francisco, CA 94109
USA
Tel: [1] (415) 474 - 3330
Tel: [1] (877) TRI - PTEL (toll-free)
Fax: [1] (415) 292-3311
Web: www.triptel.com
E-mail: rental@triptel.com

Argentina

Argentina Phone Rental
San Martin 945 6 49
Buenos Aires
Argentina
Tel: [54] (11) 4311-2933
Fax: [54] (11) 4315-2801
Web: www.phonerental.com.ar
E-mail: info@phonerental.com.ar

Australia

Cellhire
Level 4; 78 Liverpool Street
Sydney NSW 2000
Australia
Tel: [61] (0) 2 9286 9494
Fax: [61] (0) 2 9261 0707
Web: www.cellhire.com.au
E-mail: sydney@cellhire.com

Notes

Brazil

Global Presence (Nextel)
Av. Maria Coelho Aguiar
215 Block D 70 andar; Jd. São Luis
São Paulo – SP05804900 Brazil
Tel: [55] (11) 3748-1000
Web: www.nextel.com.br

China

France Telecom China
Fuhua Mansion, Bldg. B, 9th Floor
North Chaoyangmen 8,
Dongchen Avenue
Beijing 1000027 China
Tel: [86] (10) 6554-1123
Web: www.orangehk.com

France

Cellhire SA
176 Avenue Charles de Gaulle
92522 Neuilly Sur Seine CEDEX
France
Tel: [33] 1 41 43 79 40
Fax: [33] 1 40 88 04 10
E-mail: paris@cellhire.com

EuroExaphone
39 rue Saint-Ferdinand
Paris 75017 France
Tel: [33] (1) 4409-7778
Web: www.gofrance.about.com

Rent a Phone in France Online
Web: www.franceline.com/
rentaphone.htm
E-mail: corval@franceline.com

Germany

MOBILCOM
Hollerstrasse 125; Postfach 520
Rendburg-Budelsdorf; 24753 Germany
Web: www.mobilcom.de

Telekommunicationsdieste GmbH
Lyoner Strasse 9
Frankfurt am Main; 60528 Germany
Tel: [49] (69) 929-010
Web: www.equant.com

T-Mobile
Tel: [49] (180) 330 2202
Web: www.t-mobile.de

Hong Kong

Trident Telecom Ventures, Ltd.
Unit 1004, 10/F., AXA Center
151 Glouchester Road
Wanchai, Hong Kong, China
Tel: [852] 2121-1811
Web: www.hkhousewarefair.com

India

BPL Mobile Communications
BPL Center; Mahim; Mubai India
Tel: [91] (22) 432-3777; Fax: [91] (22)
431-2255

India Tours Online
Mobile Phone Rentals
Web: www.indiatoursonline.com/tools/
mobilerentals.html

Orange/Hutchison Max Telecom
Stanrose House
Standard Mills Compound
Prabhadevi; Mubai, India
Tel: [91] (22) 431-1111
Fax: [91] (22) 431-3456

Italy

WIND Telecom
48 Via Cesare Guilio Viola
Roma 00148 Italy
Tel: [39] (6) 8509-3999; Web:
www.wind.it

Equant
Via Tucidide; Milan 20134 Italy
Tel: [39] (2) 752-891; Web:
www.equant.com

Japan

Aitec Co. LTD
4-25-13 Nishimachi,
Toyota-shi Aichi
Japan 471-0025
Web: www.digi-promotion.com

Japan Handy Phone
Tourism World Inc.
402 Fukushima Dai 2 Bldg. 4F
366 Yamabuki-cho, Shinjuku-ku, Tokyo
162-0801 Japan
Tel: [81] (0)3 5225-2125
Fax: [81] (0)3 5225-2124
E-mail: sales@japanphone.com

JCR SYSTEMS
U-Building 3F, 2-16-2 Nihonbashi-
Ningyo-cho, Chuo-ku
Tokyo 103-0013 Japan
Tel: [81] (3) 5701-9200
Web: www.jcrcorp.com

Mobell Japan
Tel: [81] (0) 120 423 524
Web: www.mobell.co.jp

Sony Finance International
Tokyo International Airport
Terminal 1 B1; Narita Airport Station
Keisei Travel Service; Tokyo Japan
Tel: [81] (476) 326-391
Web: www1.sonyfinance.co.jp

Notes

Mexico

Nextel De México, S.A. de C.V.
Blvd. Manuel Avila Camacho No. 36
piso 9
Col. Lomas de Chapultepec
C.P. 11000
Mexico D.F.
Tel: [52] (5) 278-4000
Web: www.nextel.com.mx

Russia

The Russia Adventure Club
Moscow 115054
Russia
Tel: [7] (095) 233-3490; Fax: [7] (095)
233-3490
Web: www.elta.ru/travel
E-mail: vladimirf@extremetravel.info

South Africa

CelluRent
PO Box 7388
Roggebaai 8012
Cape Town, South Africa
Tel: [27] (21) 418 5656
Fax: [27] (21) 419 6735
Web: www.cellurent.co.za
E-mail: service@cellurent.co.za

GlobeCast
1 Park Road Richmond
Johannesburg
South Africa
Tel: [27] (11) 482-2790
Web: www.globecast.com

Spain

OnSpanishTime.COM
Tel/Fax: [34] (91) 547 8575
Cell: [34] 656266844

Web: www.onspanishtime.com
E-mail: jeremy@onspanishtime.com
WANADOO
Velazquez 12
Madrid 28006 Spain
Tel: [34] (91) 252-1956
Web: www.wanadoo.es

UK

Adam Phones
5 Dolphin Square
Edensor Road; London
England; UK
Tel: [1] (866) GSM-HIRE
Web: www.adamphones.com

CellPhone Rentals Limited
28 Meadowside; Abington
England UK
Tel: [44] (1235) 201-063
Web: www.rent-a-cellphone.co.uk

Hirefone
Hamilton House; 1 Temple Avenue
London
England UK
Tel: [44] (1904) 671-010
Web: www.hirefone.com

Mobell Communications Ltd.
The Winding House
Walkers Rise
Rugeley Road
Hednesford
Staffordshire
WS12 5QU
United Kingdom
Tel: 0800 24 35 24 (toll-free UK)
Tel: [44] 1543 426999 ; Fax: [44] 1543
426126
Web: www.mobell.co.uk
E-mail: website.enquiries@mobell.com

Notes

_____
_____
_____
_____
_____
_____
_____
_____
_____
_____
_____
_____

U.S. Cellular Provider Table

A common question among international travelers is, "will my cell (mobile) phone work when I get there?" The answer depends on two factors, 1) whether or not your cellular service provider has service in the particular country and 2) the type of phone you have.

The following table is for those who have service with a U.S. cellular provider and are traveling to another country.

Key

Yes = Provider has service. However, check with your provider prior to travel to find out whether you can use your existing phone or if you must rent or purchase a compatible phone.

No = Provider does not have service in the subject country.

Notes

☛ AT&T Wireless is now part of the Cingular System.

☛ If your provider has service you will still need to set up an international calling plan prior to departure.

☛ Consider the temporary rental of a compatible cell phone, including service, prior to your departure. Refer to "Cell Phone Rental Services" on page 38 for international rentals.

This information is current as of March 2005 and is constantly changing.

Country	Cingular	Nextel	Sprint	Verizon	T-Mobile
Afghanistan	No	No	No	No	Yes
Albania	Yes	Yes	Yes	Yes	Yes
Algeria	No	No	No	Yes	No
American Samoa	No	No	No	No	No
Andorra	Yes	No	Yes	Yes	Yes
Angola	No	No	No	No	No
Anguilla	Yes	No	Yes	No	Yes
Antigua & Barbuda	Yes	No	Yes	No	Yes
Argentina	Yes	Yes	Yes	No	Yes
Armenia	No	No	Yes	No	Yes
Aruba	No	No	Yes	Yes	Yes
Australia	Yes	Yes	Yes	Yes	Yes
Austria	Yes	Yes	Yes	Yes	Yes
Azerbaijan	Yes	Yes	Yes	No	Yes
Bahamas	No	No	Yes	No	Yes
Bahrain	Yes	Yes	Yes	Yes	Yes
Bangladesh	Yes	Yes	Yes	No	Yes
Barbados	Yes	Yes	Yes	No	Yes
Belarus	Yes	No	Yes	No	Yes
Belgium	Yes	Yes	Yes	Yes	Yes
Belize	Yes	Yes	No	No	Yes
Benin	No	No	No	No	No
Bermuda	Yes	No	Yes	Yes	Yes

Country	Cingular	Nextel	Sprint	Verizon	T-Mobile
Bhutan	No	No	No	No	No
Bolivia	No	No	Yes	Yes	Yes
Bosnia & Herzegovina	No	Yes	Yes	Yes	Yes
Botswana	No	Yes	Yes	Yes	Yes
Brazil	Yes	Yes	Yes	Yes	Yes
Brunei	Yes	Yes	No	Yes	Yes
Bulgaria	Yes	Yes	Yes	Yes	Yes
Burkina Faso	No	No	No	No	Yes
Burundi	No	No	No	No	No
Cambodia	Yes	Yes	Yes	Yes	Yes
Cameroon	No	Yes	Yes	Yes	Yes
Canada	Yes	Yes	Yes	Yes	Yes
Cape Verde	Yes	No	No	No	No
Cayman Islands	Yes	No	Yes	No	Yes
Central African Republic	No	No	No	No	No
Chad	No	No	No	No	Yes
Chagos (Diego Garcia)	No	No	No	No	No
Chile	Yes	No	Yes	No	Yes
China (PRC)	Yes	Yes	Yes	Yes	Yes
Colombia	No	No	No	No	No
Comoros	No	No	No	No	No
Congo, Dem. Rep. of (Kinshasa)	No	No	No	Yes	Yes
Congo, Rep. of (Brazzaville)	No	No	No	No	Yes
Cook Islands	No	No	No	No	No
Costa Rica	No	No	No	Yes	No
Côte d'Ivoire	Yes	Yes	Yes	No	No
Croatia	Yes	Yes	Yes	Yes	Yes
Cuba	No	No	No	No	No
Cyprus	Yes	Yes	Yes	Yes	Yes
Czech Republic	Yes	Yes	Yes	Yes	Yes
Denmark	Yes	Yes	Yes	Yes	Yes
Djibouti	No	No	No	No	Yes
Dominica	Yes	No	Yes	No	Yes
Dominican Republic	No	Yes	Yes	Yes	Yes
Ecuador	No	No	No	No	Yes
Egypt	Yes	Yes	Yes	Yes	Yes

Country	Cingular	Nextel	Sprint	Verizon	T-Mobile
El Salvador	No	No	No	Yes	Yes
Equatorial Guinea	No	No	No	Yes	Yes
Estonia	Yes	Yes	Yes	Yes	Yes
Ethiopia	Yes	No	No	No	No
Faeroe Islands	Yes	No	Yes	Yes	Yes
Falkland Islands	No	No	No	No	Yes
Fiji	Yes	Yes	Yes	Yes	Yes
Finland	Yes	Yes	Yes	Yes	Yes
France	Yes	Yes	Yes	Yes	Yes
French Antilles	No	No	No	No	No
French Guiana	No	Yes	Yes	Yes	Yes
French Polynesia	No	No	No	No	Yes
Gabon	No	No	Yes	Yes	Yes
Gambia	No	No	Yes	No	Yes
Georgia	No	Yes	Yes	No	Yes
Germany	Yes	Yes	Yes	Yes	Yes
Ghana	Yes	No	Yes	No	Yes
Gibraltar	Yes	Yes	Yes	Yes	Yes
Greece	Yes	Yes	Yes	Yes	Yes
Greenland	Yes	No	No	No	Yes
Grenada	Yes	No	Yes	No	Yes
Guadeloupe	No	Yes	Yes	Yes	Yes
Guam	No	No	No	No	No
Guatemala	No	No	Yes	No	Yes
Guinea	No	No	No	No	No
Guinea-Bissau	No	No	No	No	No
Guyana	No	Yes	No	No	Yes
Haiti	No	No	No	No	No
Honduras	No	No	Yes	No	Yes
Hong Kong	Yes	Yes	Yes	Yes	Yes
Hungary	Yes	Yes	Yes	Yes	Yes
Iceland	Yes	Yes	Yes	Yes	Yes
India	Yes	Yes	Yes	Yes	Yes
Indonesia	No	Yes	Yes	Yes	Yes
Iran	No	No	No	No	No
Iraq	Yes	No	No	No	Yes
Ireland	Yes	Yes	Yes	No	Yes
Israel	Yes	Yes	Yes	Yes	Yes
Italy	Yes	Yes	Yes	Yes	Yes
Jamaica	Yes	Yes	Yes	No	Yes

Country	Cingular	Nextel	Sprint	Verizon	T-Mobile
Japan	No	No	Yes	No	Yes
Jordan	Yes	Yes	Yes	Yes	Yes
Kazakhstan	Yes	Yes	Yes	No	Yes
Kenya	Yes	Yes	Yes	Yes	Yes
Korea, North	No	No	No	No	No
Korea, South	No	No	No	Yes	Yes
Kuwait	Yes	No	Yes	No	Yes
Kyrgyzstan	Yes	No	Yes	No	Yes
Laos	No	No	No	No	No
Latvia	Yes	Yes	Yes	Yes	No
Lebanon	Yes	Yes	Yes	No	Yes
Lesotho	No	No	No	No	No
Liberia	No	No	No	No	No
Libya	No	No	No	No	No
Liechtenstein	Yes	Yes	Yes	Yes	Yes
Lithuania	Yes	Yes	Yes	Yes	Yes
Luxembourg	Yes	Yes	Yes	Yes	Yes
Macau	Yes	Yes	Yes	Yes	Yes
Macedonia	Yes	Yes	Yes	Yes	Yes
Madagascar	Yes	Yes	Yes	Yes	Yes
Malawi	No	No	No	Yes	No
Malaysia	Yes	Yes	Yes	Yes	Yes
Maldives	Yes	No	No	Yes	No
Mali	No	No	No	Yes	Yes
Malta	Yes	Yes	Yes	Yes	Yes
Marshall Islands	No	No	No	No	No
Martinique	No	Yes	Yes	Yes	Yes
Mauritania	No	No	No	No	Yes
Mauritius	Yes	Yes	Yes	Yes	Yes
Mexico	No	Yes	Yes	Yes	Yes
Midway Islands	No	No	No	No	No
Moldova	Yes	No	No	No	Yes
Monaco	Yes	Yes	Yes	Yes	Yes
Mongolia	No	No	No	No	Yes
Montserrat	Yes	No	Yes	No	Yes
Morocco	Yes	Yes	Yes	Yes	Yes
Mozambique	Yes	No	Yes	Yes	Yes
Myanmar (Burma)	No	No	No	No	No
Namibia	Yes	Yes	Yes	Yes	Yes
Nepal	No	No	No	Yes	No

Country	Cingular	Nextel	Sprint	Verizon	T-Mobile
Netherlands	Yes	Yes	Yes	Yes	Yes
Netherlands Antilles	No	No	Yes	Yes	Yes
New Caledonia	No	No	No	No	No
New Zealand	Yes	Yes	Yes	Yes	Yes
Nicaragua	No	No	No	No	Yes
Niger Republic	No	No	No	No	No
Nigeria	No	No	Yes	No	Yes
Northern Mariana Islands	No	No	No	No	Yes
Norway	Yes	Yes	Yes	Yes	Yes
Oman	Yes	No	Yes	Yes	Yes
Pakistan	Yes	No	Yes	No	Yes
Palau	No	No	No	No	No
Panama	No	No	No	No	Yes
Papua New Guinea	No	No	No	No	No
Paraguay	No	No	Yes	Yes	Yes
Peru	Yes	Yes	Yes	No	Yes
Philippines	Yes	Yes	Yes	No	Yes
Poland	Yes	Yes	Yes	Yes	Yes
Portugal	Yes	Yes	Yes	Yes	Yes
Puerto Rico	Yes	No	No	Yes	No
Qatar	Yes	No	Yes	Yes	Yes
Reunion	Yes	No	Yes	Yes	Yes
Romania	Yes	Yes	Yes	Yes	Yes
Russia	Yes	Yes	Yes	Yes	Yes
Rwanda	No	No	No	No	No
St. Kitts & Nevis	Yes	No	Yes	No	Yes
St. Lucia	Yes	No	No	No	Yes
St. Vincent & Grenadines	Yes	No	Yes	No	Yes
San Marino	Yes	No	Yes	Yes	Yes
São Tomé & Principe	No	No	No	No	No
Saudi Arabia	Yes	Yes	Yes	No	Yes
Senegal	Yes	No	Yes	No	Yes
Serbia & Montenegro	No	Yes	Yes	Yes	Yes
Seychelles	Yes	Yes	Yes	Yes	Yes
Sierra Leone	No	No	No	No	No
Singapore	Yes	Yes	Yes	Yes	Yes
Slovak Republic	Yes	Yes	Yes	Yes	Yes
Slovenia	Yes	Yes	Yes	Yes	Yes

Country	Cingular	Nextel	Sprint	Verizon	T-Mobile
Solomon Islands	No	No	No	No	No
Somalia	No	No	No	No	No
South Africa	Yes	Yes	Yes	Yes	Yes
Spain	Yes	Yes	Yes	Yes	Yes
Sri Lanka	Yes	Yes	Yes	Yes	Yes
Sudan	No	No	No	No	No
Suriname	No	No	Yes	Yes	Yes
Swaziland	No	No	No	No	No
Sweden	Yes	Yes	Yes	Yes	Yes
Switzerland	Yes	Yes	Yes	Yes	Yes
Syria	No	No	No	Yes	No
Taiwan	Yes	Yes	Yes	Yes	Yes
Tajikistan	No	Yes	No	No	Yes
Tanzania	Yes	Yes	Yes	Yes	Yes
Thailand	Yes	Yes	Yes	Yes	Yes
Togo	No	Yes	No	No	No
Tonga	No	No	No	No	No
Trinidad & Tobago	No	Yes	Yes	No	Yes
Tunisia	Yes	Yes	Yes	Yes	Yes
Turkey	Yes	Yes	Yes	Yes	Yes
Turkmenistan	Yes	No	Yes	No	Yes
Turks & Caicos	Yes	No	Yes	No	Yes
Tuvalu	No	No	No	No	Yes
Uganda	Yes	Yes	Yes	Yes	Yes
Ukraine	Yes	Yes	Yes	No	Yes
United Arab Emirates	Yes	Yes	Yes	Yes	Yes
United Kingdom	Yes	Yes	Yes	Yes	Yes
United States	Yes	Yes	Yes	Yes	Yes
Uruguay	No	No	Yes	No	No
Uzbekistan	Yes	Yes	Yes	Yes	Yes
Vanuatu	No	No	No	No	No
Venezuela	Yes	Yes	Yes	Yes	Yes
Vietnam	Yes	Yes	Yes	Yes	Yes
Virgin Islands (U.K.)	No	No	No	No	Yes
Virgin Islands (U.S.)	Yes	No	Yes	No	No
Western Samoa	No	No	No	No	No
Yemen	Yes	No	No	No	Yes
Zambia	No	No	No	Yes	No
Zimbabwe	Yes	Yes	Yes	No	Yes

SMS & MMS

Cell Phone Text Messaging (SMS) & Multimedia Messaging Services (MMS)

What Is SMS?

SMS stands for Short Messaging Service and is popularly known as "text messaging." In its simplest form, SMS is simply the sending and receiving of short text messages between GSM cell phones, or between cell phones and e-mail servers. If the receiving party's cell phone is turned off, any messages received are held in a queue for later viewing; the SMS service center will retransmit them for a period of three to seven days, depending on the cellular provider. SMS messages are usually relayed abroad to wherever a user happens to be traveling.

SMS has been part of GSM network capabilities since 1992. SMS is the fastest growing method of communicating by cell phones in the world today.

SMS Messages

SMS messages are limited to alphanumeric characters associated with buttons on the cell phone. Until very recently, each message was limited to 160 Latin characters, 140 Cyrillic characters, or 70 characters in languages using non-Latin character sets (Arabic and Simplified Chinese, for example).

SMS Shorthand

Message size limitations have led to the creation of an English-based shorthand language that lets users squeeze more input into each message. This shorthand has also been driven by the young as a "hip" way to communicate. See "Text Messaging (SMS) Shorthand" on page 49 for many examples.

Benefits of SMS

The major benefit of text messaging is that it is much less expensive for the user than traditional voice calls. Transmitting text messages uses very little radio bandwidth and, unlike voice calls, cellular providers do not have to transmit text messages in real time. Thus, the charge per message is significantly less than for a voice call. Carried to an extreme, cost-conscious users in third-world countries often merely dial a cell phone number and hang up before the phone is answered as a means of conveying a "hello".

Not surprisingly, text messaging has become enormously popular among teenagers and students, who have limited funds and are comfortable with e-mailing text messages anyway. However, people of all ages and lifestyles regularly use text messaging whenever communicating about scheduling, meetings, or other precise topics. Since these messages are transmittable to e-mail servers, the receiving or transmitting parties can include anyone with an e-mail account.

The majority of cellular carriers also offer optional text-messaging alerts for information such as sports scores, stock quotes, and news headlines; these are delivered to the cell phone at regular intervals.

SMS text messaging is currently in competition with other cellular messaging options such as WAP-based e-mail, paging services, instant messaging, and short-range radio. SMS, however, is easily the front runner because of its simplicity, convenience, and price.

With the rapid development of GSM 1900 networks in North America, text messaging has recently caught on in that continent, but the high allocation of "included anytime minutes" offered by U.S. cellular providers provide little financial incentive for not using regular voice communications.

Sending and Receiving SMS

To use SMS, the user must first subscribe to SMS service through a cellular service provider. This is usually included in GSM service.

To send a message, the user presses the text message key, enters the destination phone number, inputs the message using the alpha-numeric keys on the phone and then presses the "Send" key.

When a message is received, most cell phones display a special SMS graphic. Also, some phones let the user pick a tone that is sounded when messages are received. The user then manipulates the scroll keys to view the text of the message. While the user views a message, the phone displays options to save, reply, or forward.

Differences Between Phones

There are some minor differences between phones with respect to sending text messages. The user typically enters the recipient's cell phone number or e-mail address, and then types the message using the phone's keypad. The user can transmit the message right away, or save it for later transmission.

Predictive Text Input

Some of the more modern GSM phones have "predictive text input," a smart software that will predict words before the user finishes spelling them. This feature minimizes the number of key punches. Without the predictive text input feature, for example, the user punches the number '2' key once for 'a', twice for 'b', and three times for 'c'. With the feature the phone can "learn" certain words and input the full word after only a few letters have been entered.

EMS and MMS

With the introduction of 3rd Generation (3G) cellular networks, the humble "text" in text messaging has new company. The 3G standards enable users to have access to EMS (Enhanced Messaging Services) and MMS (Multimedia Messaging Services). These extensions of SMS allow users to exchange multimedia communications (such as images, animations, audio, and video clips) through compatible phones. While still under development, the second phase of MMS technology will allow streaming video, but will require 3G technology due to unlimited size of the message. Current MMS technology uses GPRS (General Packet Radio Services) and is already included in most new GSM and CDMA phones and service agreements.

EMS is the intermediary technology that offers more capabilities than SMS, but less than MMS. EMS does not require the network infrastructure upgrades that MMS requires. The size of an MMS message is virtually unlimited. This does not mean that cellular networks won't impose their own size restrictions. Currently, Ericsson, Motorola, and Nokia all offer MMS-capable devices.

Notes

Text Messaging (SMS) Shorthand

AAMOF .... As a Matter of Fact	FYI .... For Your Information
ABT .... About	F2F .... Face to Face
ADN .... Any Day Now	F2T .... Free to Talk
AFAIK .... As Far As I Know	GAL .... Get A Life
AFK .... Away From Keyboard	GBF .... Great Big Hug
AKA .... Also Known As	GF .... Girlfriend
ASAP .... As Soon As Possible	GFC .... Going for Coffee
ATB .... All the Best	GIVZA .... Give Us A
ATK .... At The Keyboard	G2G .... Got to Go
ATM .... At The Moment	GMTA ... Great Minds Think Alike
AYPI .... And Your Point Is?	GONA .... Going to
A3 .... Anytime, Anywhere, Any-place	GR8 .... Great!
B .... Be	GUDAM .... Good Morning
BAK .... Back At Keyboard	GUDNITE .... Good Night
BAW .... Bells and Whistles	G9 .... Genius
BBL .... Be Back Later	HAND .... Have a Nice Day
BBS .... Be Back Soon	HIH .... Hope It Helps
BCNU .... Be Seeing You	HRU .... How Are You?
BF .... Boyfriend	HSIK .... How Should I Know?
BFN/B4N .... Bye For Now	HTH .... Hope This Helps
BION .... Believe It Or Not	HV .... Have
BN .... Being	H8 .... Hate
BRB .... Be Right Back	IAC .... In Any Case
BRT .... Be Right There	IC .... I See
BTW .... By The Way	IGTP .... I Get the Point
BWD .... Backward	ILU .... I Love You
B2B .... Back to Back	IMHO ...... In My Humble Opinion
B4 .... Before	IMO .... In My Opinion
B4N .... Bye For Now	IOW .... In Other Words
C .... See	IRL .... In Real Life
CU .... See You	IUSS .... If You Say So
CUL8R .... See You Later	JAM .... Just a Minute
CWOT . Complete Waste of Time	JIC .... Just In Case
CYA .... Cover Your Ass	JK .... Just Kidding
CZIN .... Season	JTLYK .... Just to Let You Know
D .... The	K .... Okay
DYK .... Do You Know	KHYF .... Know How You Feel
F .... If	KISS .... Keep It Simple Stupid
FC .... Fingers Crossed	LDR Long Distance Relationship
FOAF .... Friend of a Friend	LL&P ...... Live Long and Prosper
FWIW .... For What It's Worth	LMAO .... Laugh My Ass Off
FYEO .... For Your Eyes Only	LNK .... Love and Kisses
	LO .... Hello

LOL	Laughing Out Loud
LTNS	Long Time No See
LUVU	Love You
L8R	Later
MHOTY	My Hat's Off to You
MOM	Moment
MTE	My Thoughts Exactly
MTNG	Meeting
M8	Mate
NE1	Anyone
NOYB	None of Your Business
NRN	No Reply Necessary
NUTN	Nothing
OFIS	Office
OIC	Oh I See
OTOH	On the Other Hand
PCM	Please Call Me
PITA	Pain In The Ass
PLS	Please
PMJI	Pardon Me for Jumping In
PPL	People
PRT	Party
PRW	Parents Are Watching
Q	Queue
QPSA?	Que Pasa?
R	Are
ROFL	Rolling On The Floor Laughing
RUOK	Are You OK?
S	Is
SK8	Skate
SME1	Someone
STATS	Your sex and age
SUM1	Someone
ASL	Age, Sex, Location
THX	Thank You
TMI	Too Much Information
TTFN	Ta-Ta For Now!
TTYL	Talk To You Later
U	You
UR	You are
URT1	You Are the One
U2	You Too
U4E	Yours For Ever
W	With, Were, Where
WAN2	Want to
WB	Welcome Back
WENJA	When Do You
WERJA	Where Do You
WK	Week
WKND	Weekend
W/O	Without
WRU	Where Are You?
WTF	What The Fuck
WTG	Way To Go!
WUF	Where Are You From?
WUWH	Wish You Were Here
W8	Wait
W8AM	Wait a Minute
XFER	Transfer
XLNT	Excellent
YBS	You'll Be Sorry
2	to, too, two
2DAY	Today
2MORO	Tomorrow
2NITE	Tonight
4	for, four
4EVER	Forever
:-)	Smiley
(-:	Also smiling
:)	Smiling without a nose
:')	Happy and crying
:-( )	Smiling with mouth open
8-)	Smiling with glasses
[:-)	Smiling with walkman
:-)8	Smiling with bow tie
{:-)	Smiling with hair
d:-)	Smiling with cap
C|:-)	Smiling with top hat
(:-)	Smiling with helmet
:-)=	Smiling with a beard
&:-)	Smiling with curls
#:-)	Smiling with a fur hat
:-D	Laughter
;-)	Twinkle
;)	Twinkle, without nose
:-*	Kiss
@}--\-,---	A rose
:-(	Sad
:(	Sad, without nose
:'-(	Crying
:-c	Unhappy
O :-)	An angel
:-9	Salivating

Satellite Telephones

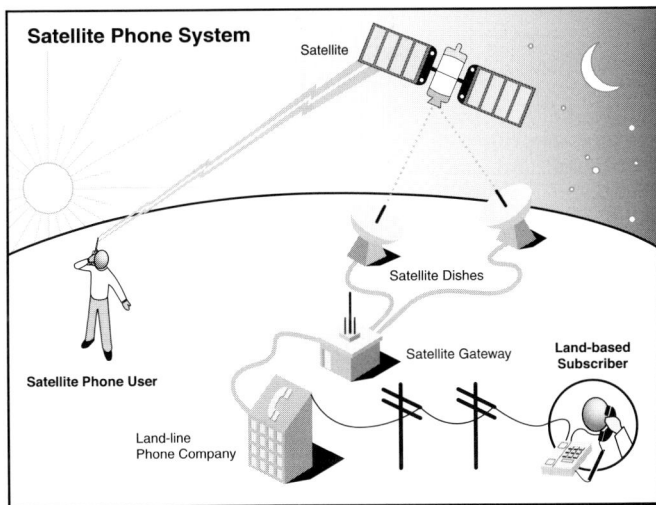

Satellite Phone System

Satellite

Satellite Dishes

Satellite Gateway

Land-based
Subscriber

Satellite Phone User

Land-line
Phone Company

What Is a Satellite Telephone?

Satellite telephones are radio devices that send and receive communication signals (voice and data) via man-made satellites launched into Earth orbit by various public and private companies and governments.

A satellite telephone can be equipped to send and receive voice, fax, or computer data. Some are also used to connect to the Internet and send and receive e-mail.

The single greatest advantage of a satellite telephone is that it can send and receive signals (calls) from virtually anywhere in the world. Thus, these telephones are indispensable in maritime transport and navigation as well as for use by anyone working in remote, uninhabited or slightly inhabited regions of the world where land-line telephone services and cellular networks do not reach.

Who Uses Them?

Satellite telephones are an essential communications link for businesspeople, explorers, geologists, mining companies, diplomats, journalists, mariners, the military and even back-country hikers where conventional cellular or land-based systems are unavailable, unreliable, or in risk of destruction.

Satellite telephones and airtime are considerably more expensive than traditional cellular or land-line phones and are therefore not as widely used. Demand has not met providers' expectations and several providers have had financial difficulty over the past few years.

Different types of satellite phones are available for different needs. Marine equipment and vehicular telephones are available as well as fixed site and portable systems.

How Do They Work?

Essentially, satellite telephones are similar to cellular telephones in that radio signals are transmitted and decoded. The Globalstar Communications System, for example, uses digital cellular CDMA technology.

The Phone When you turn on your satellite phone and dial a number, the phone converts your voice or data into a radio signal that is transmitted through its oversize antenna to a satellite in Earth orbit.

The Antenna In some systems, the radio signal is sent through a directional antenna that focuses the signal from the handset toward a satellite through means of a tracking device in the phone. In such systems, a call must be placed from a stationary terminal; that is, the caller cannot be mobile while transmitting. Other systems operate with a non-directional antenna, eliminating the need for the internal tracking device.

The Satellite Each phone network has a number of Earth-orbit satellites. Although these are marvels of engineering, they fundamentally contain an antenna, transceiver (for receiving and sending radio signals), and a power supply.

The satellite receiving the signal may be in a geostationary (synchronous) orbit, meaning that it remains in orbit over a specific geographic area at 35,600 km above the Earth. Telephone calls using such satellites have noticeable voice delays and echo effects of approximately one quarter of a second.

Qualcomm GSP 1600 Globalstar / CDMA / AMPS Tri-Mode (Satellite) Phone. Note large antenna.

Alternatively, the satellite may be in a lower, asynchronous (not geostationary) orbit. If you are transmitting from a fixed position, your call may be switched from satellite to satellite as each comes into and then goes out of range. Compare this to an Earth-based cellular system where the user goes into and out of range of fixed-position base stations. Telephone calls using such lower-orbiting asynchronous satellites have almost no perceptible voice delays and echo effects. Once the satellite receives a signal, it processes it for retransmission to an Earth-based satellite dish.

The Earth-based Satellite Dish The satellite's signal is picked up by one of a number of strategically located earth-based satellite dishes. The signal is once again transmitted to a satellite gateway.

The Satellite Gateway The satellite gateway receives signals from a number of different satellite dishes. The signal is forwarded to a land-line phone company and charges are assessed to the user.

The Land-line Phone Company It is important to note that satellite phone systems are distinct from local telephone services such as AT&T in the U.S. and British Telecom in the U.K. Once the signal leaves the satellite gateway it is picked up by one of these companies and "delivered" to the land-based subscriber.

How Well Do They Work?

If you are considering using a satellite phone, know the limits of the technology. These limits include: 1) geographic coverage; 2) transmission and reception quality; 3) equipment; 4) special equipment for data and fax transmission; and 5) cost.

Geographic Coverage While global coverage is often advertised, transmission and reception in some places on Earth is limited because satellite coverage is poor. In some remote areas, only certain satellite systems have coverage and there are still a few areas without any coverage. If the satellites that your phone is made to work with are geostationary, then there will be no coverage of very northern or very southern latitudes. Additionally, each such satellite can only "see" a limited part of the earth, which is the only region that can be serviced.

Your handset cannot transmit or receive if there is no satellite in range. Similarly, regional systems (e.g., MobileSat) often allow you to call out to anywhere in the world through arrangements with affiliate companies, but the handset will not transmit unless it is located within the regional satellite coverage area.

Other global systems require transmission through earth-bound stations, and therefore do not work unless an Earth station is present in your country or area. Also, some satellite systems do not have "landing rights" in some countries and try to deny service to users in those countries as a means of complying with the absence of such "landing rights."

For most communications needs, the cost of a satellite telephone may be unwarranted because less expensive and more reliable earth-bound systems are available. However, if you are on a North Sea oil platform, at a remote mine in Chile, or on a container ship in the middle of the ocean, it's the only way to go, short of setting up elaborate short-wave radios and antennas.

Transmission and Reception The quality of transmission and reception is dependent on many, often uncontrollable, factors such as weather, location of the satellite in relation to the handset, peak hours of demand, and clear view of the sky. You need to be patient with delays in transmission and understanding when signal power is insufficient; this technology is sending signals 35,000 km into orbit, not merely 5 to 10 km to a cellular base station. For laptop-like (or larger) systems, calls must be made only while the caller is stationary with the antenna directed at the satellite, whereas for handheld units, the user can be in motion.

What Equipment Is Available?

Satellite phones come in a variety of configurations. These include hand-held, portable (briefcase size), stationary, marine and aviation models. Refer to the table of equipment manufacturers on the following pages.

Satellite Telephone Networks

Refer to the tables on the following pages for information on the various satellite network providers.

What's in the Future?

Fast-paced technology is improving the access to and the quality of satellite links. The industry is highly competitive, with service providers finding new ways to shrink handset and terminal sizes, boost signals, and enhance global coverage.

Most systems remain limited by geographical coverage of the satellites or land link stations, but as competition forces providers to improve their equipment and enhance their technologies, users will continue to benefit.

Telephones that combine cellular and satellite technologies, allowing the caller to use either system, are becoming more available, and a few companies now offer global satellite computer and video network capabilities that can be specially tailored to a client's needs. Such systems are likely to become more common among global firms as video-conferencing becomes more popular.

What Do They Cost?

Hardware The cost of satellite phones ranges from US$500 to $2,000 for handheld models, and from US$2,500 to $10,000 for fixed-site equipment. Marine and commercial grade stationary units cost upwards of US$20,000.

Call Service Plan Most providers offer service plans that include a set number of minutes of airtime each month with reduced rates for off-peak time.

Airtime Airtime remains expensive and is the greatest obstacle to widespread use of this technology. Current charges range from US$1 to $12 per minute.

Some systems charge only the user of the satellite telephone, while others charge both the satellite telephone user and the caller who is transmitting to the satellite telephone.

How Do You Buy or Rent One?

While there are retail outlets for satellite phones, most are purchased or rented from specialty firms that concentrate on this technology. See "Satellite Phone Rental Companies" on page 59. See also "Satellite Phone Equipment Manufacturers" on page 57.

SATELLITE NETWORKS WORLDWIDE			
Company / Contact	System type	Coverage	Services/equipment
ACeS (ASIA Cellular Satellite) Paragon Tower #15-04/06 290 Orchard Road, Singapore 238859 Tel: [65] 6735-6656 Web: www.acesinternational.com	geostationary	Central, East, and Southeast Asia	ACeS R190 dual-mode mobile phone (GSM/satellite) ACeS FR-190 fixed user terminal ACeS aicast, high-speed data terminal ACeS Maritime Communication System (M-CAT)
Inmarsat 99 City Road London, EC1Y 1AX United Kingdom Tel: [44] (20) 7728 1777 Web: www.inmarsat.com	geostationary	worldwide	Inmarsat-Mr/GAN Inmarsat-B Transportable Inmarsat M4 Mini-M Transportable Mini-M Vehicular Inmarsat-C Inmarsat Aeronautical Inmarsat Marine
Iridium 8440 South River Pky. Tempe, AZ 85284 USA Tel: [1] (866) 947-4348 (toll-free in the USA) Tel: [1] (480) 752-5155 Web: www.iridium.com	low-earth orbit	worldwide	Iridium Series 9500 Iridium Motorola 9520 Mobile Phone Iridium Motorola 9570 Portable Dock Iridium Motorola 9501 Pager
GlobalStar 3200 Zanker Road, Bldg. 260 San Jose, CA 95134 USA Tel: [1] (877) 466-9467 (toll-free in the USA) Tel: [1] (408) 933-4000 www.globalstar.com	low-earth orbit; earth-bound relay station	worldwide; transmission requires earth station within 1,500 km of telephone	GSP 1600 Satellite Phone GCK 1400 Hands-Free Car Kit GSP 2800/2900 Fixed Satellite Phone Portable Marine Kit
MSat Mobile Satellite Ventures 10802 Parkridge Blvd. Reston, VA 20191 USA Tel: [1] (800) 405-6543 Web: www.msvlp.com	geostationary	N. America Hawaii Caribbean Northern S. America	Mitsubishi ST211 - Land Mobile ST221M - Fixed Site ST251 - OmniQuest Westinghouse Mobile and Fixed Terminals
Orbitel 21839 Atlantic Blvd. Dulles, VA 20166 USA Tel: (703) 948-8600 Fax: (703) 948-8526 Web: www.orbital.com	low-earth orbit; geosynchronous earth orbit	customized coverage	Commercial space systems Planetary spacecraft Space systems support GPS satellite navigation & earth communications STAR geo satellites
Thuraya Thuraya Satellite Telecommunications Co. P.O. Box 33344 Abu Dhabi - UAE Tel: [971] (2) 6422222 Fax: [971] (2) 6418440 Web: www.thuraya.com	geostationary	Europe North and Central Africa Middle East CIS countries South Asia	Hughes Thuraya Phone Ascom Thuraya Phone

Satellite Videophones

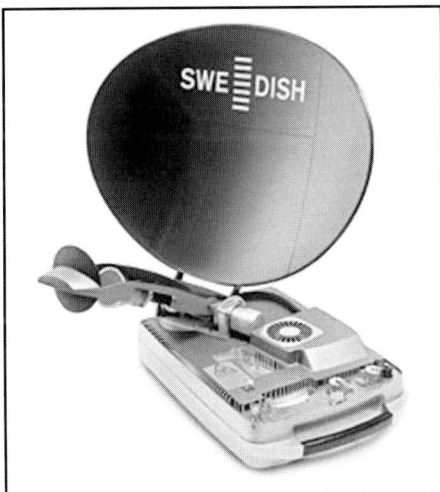

What Is a Satellite Videophone?

Satellite videophones are satellite telephones adapted for easy digital input and transmission of a video data feed in addition to audio.

Most people associate the satellite videophone exclusively with the choppy, pause-filled television news reports seen daily from more and more hot spots around the world. In the past decade, improvements in technology and portability and lower prices for both telephones and satellite services have led to explosive growth in the use of satellite videophone systems in many other activities and services, both commercial and non-commercial.

In the Past

Prior to the availability of portable systems, sending live video and audio from remote locations necessitated a satellite uplink facility (consisting of several thousand pounds of recording and transmission hardware), as well as a crew of three or four, usually working from a van. In contrast, the new portable systems are transportable and usable by a single person. The videophone terminal (the heart of the system) can be attached to practically any video camera.

System Advantages

Satellite videophones can be a godsend in any critical activity where there is no cellular or landline communications infrastructure.

First-generation portable videophone hardware cost well over US$50,000 and fitted into two suitcases weighing more than 50 kilos (110 lbs). The latest-generation hardware now sells in the US$2000 to $3000 range and weighs less than five kilos (11 lbs).

The evolution of cellular technology enables videophone systems to simultaneously send and receive other data in addition to video and audio.

For example, the videophone user can query database-driven Web sites. Software connected to these sites can analyze data and send results back via satellite to the videophone system in the remote location.

Finally, the development of broadband hardware and transmission capabilities promises a much improved video and audio quality as well as vastly greater transmission of non-video/audio data.

How Simple Is Simple?

The steps typically taken by a reporter, for example, to use a satellite videophone system are simple:

1. Unfold the Inmarsat antenna.
2. Open the videophone box.
3. Connect the camera to the videophone box.
4. Place the call.
5. Transmit video and audio data.

A single videophone terminal transmits data at 64 Kbps. A common technique to improve video and audio quality is to connect two terminals together, for a 128 Kbps per second transmission rate. Inmarsat satellite transmissions at the current 64 Kbps average around US$7 per minute. Work is being done on technology that will be able to bind several 64 kbps transmissions per terminal into a single, faster two-way stream.

Uses of Modern Satellite Videophones

1. **Video Conferencing** Two-way video-conferencing simply requires videophone systems on each end of the transmission.
2. **News Media** Reporters in the field can file real time reports from anywhere in the world.
3. **Ambulances** Paramedics transporting the injured to medical facilities in ambulances can send video images of wounds directly to emergency room staff. At the same time they can send other data such as pulse rates and blood pressure. This permits the emergency room staff to formulate preliminary analysis by the time the patient arrives, which in some critical situations can mean the difference between life and death.
4. **Medical** Medical surgeries in remote locations can be monitored by and receive input from medical experts located anywhere in the world. (A recent example was an emergency surgery in the Dominican Republic that was connected by satellite videophone to specialty surgeons at Yale University in the U.S.)
5. **Disaster Relief** Disaster aid workers can send real time images of disaster situations for analysis by experts at central locations. This is especially useful when other telecommunications infrastructures have been damaged or destroyed by natural disasters or man-made catastrophes.
6. **Military** Combat photographers can send real time reports of battle conditions to commanders.
7. **Mining/Petroleum** In the mining and petroleum industries, use of satellite video feeds has proven as effective as, and substantially less expensive than, transporting engineers into remote locations for many types of maintenance and fixes.
8. **Ecological Surveillance** Ecologists can set up a satellite videophone to monitor water supplies or pollution run-off. The system can also be triggered with a motion-detector.
9. **Meteorology** Meteorologists can track dangerous weather systems in real time.

Notes

Satellite Phone Equipment Manufacturers

Company / Contact Data	Network	Euipment
EMS Satcom Tewkesbury Gloucestershire, GL20 8HD United Kingdom Tel: [44] (0) 1684 290 020 1725 Woodward Dr. Ottawa, ON K2C 0P9 Tel: 1-800-600-9759 or [1] (613) 727-1771 Web: www.otter.co.uk	Inmarsat	EMS Storm Satphone (M4
GlobalStar 3200 Zanker Road, Bldg 260 San Jose, CA 95134 USA Tel: [1] (877) 728-7466 (toll-free in USA) Tel: [1] (408) 933-4000 Web: www.globalstar.com	Global-Star	GSP1600 Satellite Phone GCK 1410 Car Kit GSP 2800/2900 Fixed Phone
Mitsubishi Electric Americas Corporate Office 5665 Plaza Drive P.O. Box 6007 Cypress, CA 90630 USA Tel: [1] (714) 220-2500 Fax: [1] (714) 229-3854 Web: www.mitsubishielectric-usa.com	MSat	ST211 - Land Mobile MT ST221M - Fixed Site MT ST251 - OmniQuest - Transportable
Motorola 1475 W. Shure Drive Arlington Heights, IL 60004 USA Tel: 1-866-BUY-MOTO Web: www.motorola.com	Iridium	Iridium 9500 handset Iridium 9505 handset
Nera - Donegal Holdings, Ltd. 12674 Goar Rd. Houston, TX 77077-2328 USA Tel: [1] (281) 556-8886 Fax: [1] (281) 556-9573 Web: www.donegal-holdings.com/nera.htm E-mail: Sales@donegal-holdings.com	Inmarsat	World Communicator (M4) Saturn Transporter (B) WorldPhone (Mini-M) WorldPhone Voyager (Vehicular) WorldPhone Provident (high-gain antenna) Saturn Bm (B) Marine WorldPhone Marine (Mini-M)

Company / Contact Data	Network	Euipment
Thrane & Thrane Lundtoftegardsvej 93 D DK-2800 Kgs. Lyngby Denmark Tel: [45] (39) 55 88 00 Web: www.tt.dk E-mail: info@tt.dk 509 Viking Dr., Suites K, L, and M U.S. Office Virginia Beach, VA 23452 Tel: [1] (757) 463-9557 Fax: [1] (575) 463-9581	Inmarsat	TT-3080A Capsat Messenger (M4) TT-3060A Capsat (Mini-M) TT-3066-A (Mini-M) TT-3062A Capsat Rod Phone (vehicular) TT-3062D Capsat Compact Car-phone TT-3022A C/GPS (C) Land Mobile Capsat TT-3000M Aero-M TT-5000 Aero-I; TT-3024 Aero-C TT-3084A Capsat Fleet 77 (M4) (marine) TT-3064A Maritime (Mini-M) TT-3020B C/GPS Maritime Cap-sat (C) TT-3020C C/GPS GMDSS Mari-time Capsat (C) TT-3026LM eCTrack (C)
Thuraya Satellite Telecommu-nications Co. Abu Dhabi - UAE P.O. Box 33344 Tel: +971 2 6422411 Fax: +971 2 6420012 Web: www.thuraya.com	Thuraya	Hughes 7100 Hughes 7101
Westinghouse 1340 Charwood Road, Suite 1 Hanover, MD 21076 USA Tel: [1] (800) 851-4807 Web:/www.remotesatellite.com/products/wstinhsruggedtalk2.htm	MSat	RuggedTalk (transportable) FleetTalk (vehicular) WaveTalk (maritime) F1000S, L RemoteTalk (fixed site)

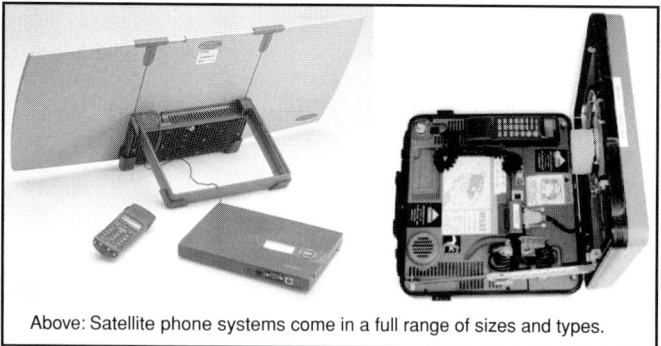

Above: Satellite phone systems come in a full range of sizes and types.

Satellite Phone Rental Companies

SATELLITE PHONE RENTALS AUSTRALIA				
Company / Contact Data	Iridium	Inmarsat	Global-Star	Other
Cellhire Pty Level 4, 78 Liverpool Street Sydney, NSW 2000 Australia Tel: [61] (8) 9286-9494; Tel: 1-866-452-7368 (toll-free U.S.) Web: www.cellhire.com	X			X (Thuraya)

SATELLITE PHONE RENTALS EUROPE				
Company / Contact Data	Iridium	Inmarsat	Global-Star	Other
3rd Planet Connections Ltd. www.3pc.co.uk	X	X		Thuraya ACeS
Callmonitor Ltd 5th Floor, 150 Regent Street London W1B 5SJ, UK UK Freephone: 0800 747 6747 USA Toll-free: 1 (800) 418-8400 International: [44] (20) 7292-9209 Web: www.satphone.co.uk	X	X	X	Thuraya
Cellhire Pty Park House, Clifton Park York, Y030 5PB, United Kingdom Tel: 0800-610-610 (toll free in U.K.) Tel: [44] 1904 610-610 Web: www.cellhire.co.uk	X	X		Thuraya Mini-M
Rent@MobilePhone 191-192 Temple Chambers, London EC4Y 0DB, UK. Tel: [44] (0)20 7353 7705; [44] 870-7500-770 Fax: [44] (0)20 7353 7706 www.rent-mobile-phone.com/satellite E-mail: hire@rent-mobile-phone.com	X	X		
Stargate Satellite Communications Ewenny House, Upper Mayland, Chelmsford, Essex, CM3 6EE, UK Tel: [44] (0)1621 773-755 Fax: [44] (0)1621 773-838 Web: www.stargate3.co.uk E-mail: StarGate3@btinternet.com	X	X	X	Thuraya ACeS

SATELLITE PHONE RENTALS NORTH AMERICA

Company / Contact Data	Iridium	Inmarsat	Global-Star	Other
Action Rent a Phone One Embarcadero Center #4100 San Francisco, CA 94111 USA Tel: [1] (415) 929-0400 Tel: [1] (800) 727-0600 (USA toll-free) Web: www.globalphone.net		X		
Anywhere Phone Rental Tel: (US toll-free) [1] (800) 768-9936 Web: www.anywhere-phone-rental.com info@anywhere-phone-rental.com North America, South America, Europe, Asia, Oceania, Africa, Middle East, Caribbean, Central America		X	X	
Crystal Communications 4814 West Commercial Blvd. Tamarac, Florida 33319 USA Tel: [1] (954) 739-2422 (International) Tel: [1] (800) 513-2422 (North America) Web: www.crystalcommunications.net info@crystalcommunications.net	X	X	X	
Hello Anywhere 67 Berkeley Street, Toronto Toronto, ON, Canada M5A 2W5 Tel: [1] 416-367-4355 Tel: [1] (888) 729-4355 (toll-free in USA) Web: www.helloanywhere.com E-mail: sales@helloanywhere.com	X	X		
Outfitter Satellite, Inc. 107 Music City Circle, Suite 101 Nashville, TN 37214 USA Tel: [1] (877) 436-2255 (U.S. toll-free) Tel: [1] (615) 695-2537 Web: www.outfittersatellite.com/	X	X	X	X (Thuraya)
RentCell 2625 Piedmont Road Ste. 56-170 Atlanta, GA 30324 USA Tel: [1] (800) 404-3093 Tel: 404-467-4508 Web: www.rentcell.com		X		
Satellite-phones.org Web: www.satellite-phones.org	X	X	X	
Sat Phone Store Atlantic Radio Telephone, Inc. 2495 NW 35th Ave. Miami, FL 33142 Tel: 866-633-9636 Tel: 1-800-940-9636 (toll-free in U.S.); [1] (305) 633-9636 Web: www.satphonestore.com/rentals/ Rentals.htm	X	X	X	X (Thuraya, Regional BGAN)

SATELLITE PHONE RENTALS
CENTRAL/SOUTH AMERICA

Company / Contact Data	Iridium	Inmarsat	Global-Star	Other
Abroad Phone Rental Web: www.abroad-phone-rental.com info@abroad-phone-rental.com Rental/service areas: Argentina, Aruba, Bolivia, Brazil, Chile, Colombia, Ecuador, French Guiana, Guyana, Paraguay, Peru, Suriname, Uruguay, Venezuela	X	X	X	
Altel Cellphone Rental Av. Cordoba 417 1 - 4 (1054) Buenos Aires Argentina Tel: [1] (800) 768-9936 (toll-free U.S.) Tel: [54] (11) 4311-5000 (Argentina) Web: www.altelphonerental.com	X			
Anywhere Phone Rental Tel: (US toll-free) [1] (800) 768-9936 Web: www.anywhere-phone-rental.com E-mail: info@anywhere-phone-rental.com	X		X	
Global Cellular Rental 501 Fifth Ave. New York, NY 10017 Tel: 1800-931-9773 Web: www.globalcellularrental.com/ cell-phone-rental-services-satellite- rental.php Email: info@globalcellularrental.com	X	X	X	

Notes

_____

_____

_____

_____

_____

_____

_____

_____

_____

_____

ACeS
Satellite Coverage Map

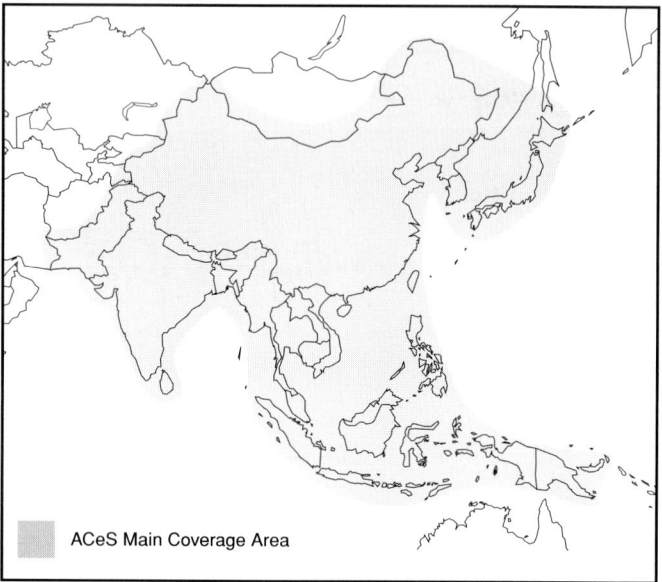

ACeS Main Coverage Area

ACeS Satellite Network Coverage

The map above illustrates satellite voice and data coverage for ACeS customers.

ACeS offers extensive satellite service in Asia, covering an area of over 11 million square miles. ACeS provides the user two types of terminals: handheld and fixed. The company's gateways provide the primary interface between ACeS and public switched telephone networks and public land mobile networks. The ACeS system uses a Garuda geostationary satellite for standard GSM services.

Corporate Headquarters

ACeS
ASIA Cellular Satellite Singapore Pte.Ltd.
Paragon Tower #15-04/06
290 Orchard Road
Singapore 238859
Tel: [65] 6735-6656; Fax: [65] 6735-6676
Web: www.acesinternational.com

GlobalStar
Satellite Coverage Map

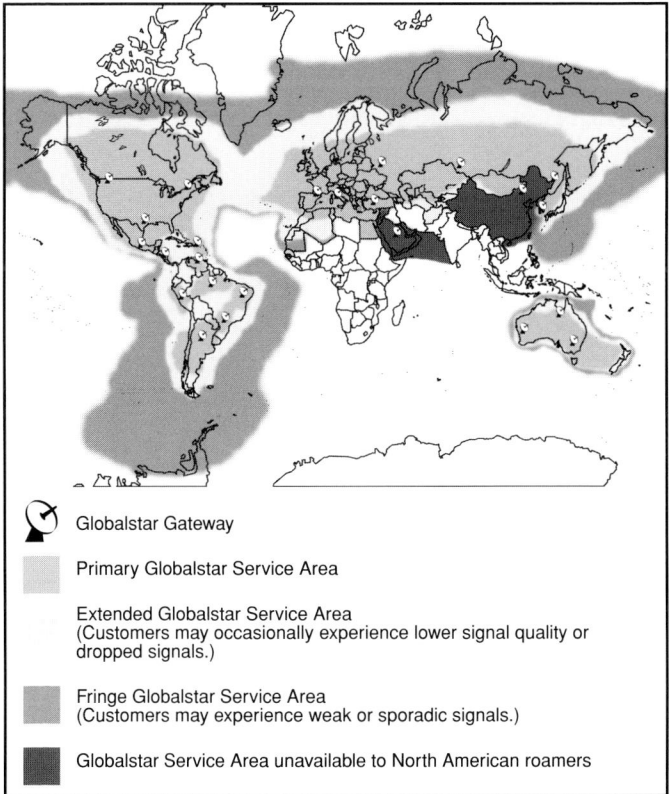

Globalstar Gateway

Primary Globalstar Service Area

Extended Globalstar Service Area
(Customers may occasionally experience lower signal quality or dropped signals.)

Fringe Globalstar Service Area
(Customers may experience weak or sporadic signals.)

Globalstar Service Area unavailable to North American roamers

GlobalStar Satellite Network Coverage

The map above illustrates current and expected future roaming satellite voice and Internet (data) access coverage available for Globalstar USA and Caribbean customers. Actual service may differ depending upon gateway development, licensing and other factors.

All service coverage areas outside GlobalStar's USA current coverage area, as shown in the map above, are subject to Globalstar USA's implementing roaming service with Globalstar Service providers.

Corporate Headquarters

Globalstar
461 S. Milpitas Blvd.
Milpitas, CA 95035 USA
Tel: [1] (408) 933-4000
Toll Free (USA): 1 (877) 728-7466; Fax: [1] (408) 933-4100
Web: www.globalstar.com; Email: sales@globalstar.com

Inmarsat
Satellite Coverage Map

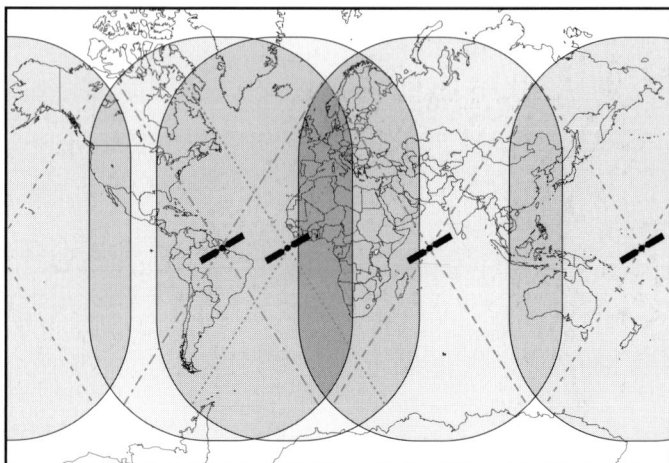

Inmarsat Satellite Network Coverage

The map above illustrates satellite voice and Internet (data) access coverage available for Inmarsat customers.

As the world's first global mobile satellite communications operator, INMARSAT offers advanced coverage for maritime, aeronautical, and land-mobile users. INMARSAT now supports links for phone, fax and data communications at up to 64kbit/s to more than 250,000 ship, vehicle, aircraft and portable terminals. Inmarsat's goal is to launch the I-4 satellite system, which from 2005 will support the Inmarsat Broadband Global Area Network (B-GAN) - mobile data communications at up to 432kbit/s for Internet access, mobile multimedia and many other advanced applications.

Corporate Headquarters

Inmarsat
99 City Road,
London, EC1Y 1AX
UNITED KINGDOM
Tel: [44] (20) 7728-1000; Fax: [44] (20) 7728-1044
Tel: [44] (20) 7728-1777 (Customer Service)
Fax: [44] (20) 7728-1142
Web: www.inmarsat.com
North America
110 Wilson Blvd., Suite 1425; Arlington, VA 22209
Tel: [1] (703) 647-4760
Fax: [1] (703) 647-4761

Iridium
Satellite Coverage Map

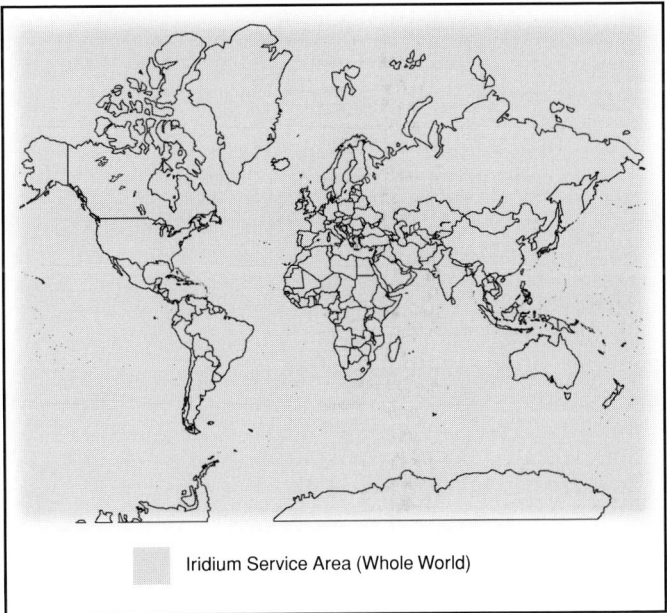

Iridium Service Area (Whole World)

Iridium Satellite Network Coverage

The Iridium Satellite System is a global, mobile satellite voice and data solution with complete coverage of the Earth (including oceans, airways and polar regions).

The Iridium solution operates through a constellation of 66 low-earth orbiting satellites operated by the Boeing Corporation.

Corporate Headquarters

Iridium
6701 Democracy Blvd., Ste. 500
Bethseda, MD 20817 USA
Tel: [1] (301) 571-6200
Fax: [1] (301) 571-6250
Sales Questions: [1] (480) 752-5155 (option 1)
Toll Free: 1-866-947-4348
Web: www.iridium.com
Email: sales@iridium.com

Business Operations

Iridium
8440 South River Parkway
Tempe, AZ 85284 USA
Tel: [1] (480) 752-1100
Toll Free (USA): 1 (866) 947-4348
General questions: info@iridium.com

MSat
Satellite Coverage Map

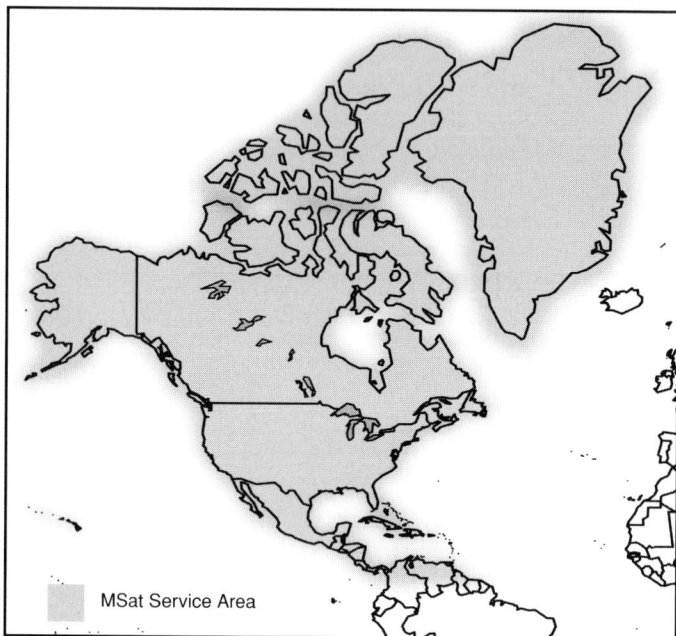

MSat Service Area

MSat Satellite Network Coverage

The map above illustrates North American satellite voice and data coverage available for MSat customers. The North American mobile satellite system (MSAT) began providing the United States and Canada with mobile satellite services in 1995.

The MSat system provides three generic service types: Mobile Telephone (MTS); Mobile Radio - Base Station operations (MRS); and Mobile Data. MSAT is the first dedicated system in North America for mobile telephone, radio, facsimile, paging, position location, and data communications for users on land, at sea, and in the air.

Corporate Headquarters Canada

MSV Network / MSat (Canada)
1601 Telesat Court
Ottawa, ON K1B 1B9 Canada
Tel: (800) 216-6728; Fax: (703) 742-4120
Customer Service: 1-800-216-6728
Web: www.msvlp.com

Corporate Headquarters USA

MSV Network / MSat (USA)
10802 Parkridge Boulevard
Reston, VA 20191 USA
Tel: +1 (877) 678-2920; Fax: (703) 390-2770
Web: www.msvlp.com; E-mail: info@msvlp.com

Thuraya
Satellite Coverage Map

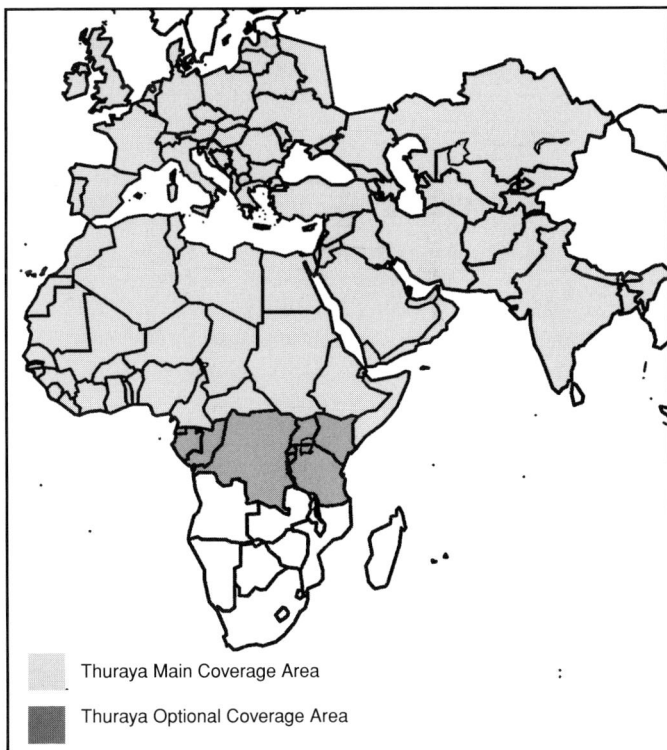

Thuraya Main Coverage Area

Thuraya Optional Coverage Area

Thuraya Satellite Network Coverage

The map above illustrates satellite voice and data coverage for Thuraya customers. Thuraya offers dual-mode terminals that offer satellite and GSM connectivity. This enables users to switch between less expensive GSM cellular coverage and satellite coverage in areas that are outside the terrestrial system.

Corporate Headquarters

Thuraya Satellite Telecommunications Co.
Abu Dhabi, UAE
PO Box 33344
Tel: [971] (2) 642-2222; Fax: [971] (2) 641-8440
Web: www.thuraya.com

Customer Service Center

Tel: [971] (2) 642-2411
Fax: [971] (2) 642-0012
Subscribers: helpline@thuraya.com
General Inquiries: general@thuraya.com
Contracts, tenders, and legal issues:
legal_contracts@thuraya.com

The Internet

What Is the Internet?

The Internet is the world's largest, fastest-growing and most important computer network. In simple terms, the Internet is millions of computers that are linked together so that a person sitting at a computer in one location can access data or communicate with a person on a computer at another location.

The word "Internet" actually means "interconnected network of networks." Therefore, it is not just a network of computers, but a network of networks.

The Internet started out as a US Department of Defence project and was called the ARPANET (Advanced Research Projects Agency Network). In 1969, the network linked four nodes: the University of California at Los Angeles, SRI (Stanford Research Institute at Stanford), the University of California at Santa Barbara and the University of Utah. In 1972, e-mail was invented. There are now tens of millions of nodes and the Internet continues to grow.

For definition of terms refer to "Glossary" on page 149.

How Does It Work?

Your Computer For the user it starts with having access to a computer with the right software (a browser), a data connection (a dial-up modem, cable modem, DSL, ISDN, T1 or other communications line) and connection to the Internet itself through an Internet Service Provider (ISP) or online service.

The Browser A "browser" is an application software that enables the user to access the Internet and the World Wide Web. All browsers have an address window where the user types in an Internet address (Web address or URL), clicks "GO" and the requested Web site opens in the browser's main window. The browser enables the user to navigate through a Web site by clicking on hyperlinked words or images. There are many browsers, but the most popular are Microsoft's Internet Explorer and Netscape Navigator.

The Data Connection Your computer needs a data connection to the Internet, or more precisely, to your ISP, which actually connects you to the Internet. The most common data connection is a modem which is a small hardware device installed in your computer or between your computer and the outside data line. The modem converts your computer's digital pulses to audio that can be transmitted over a telephone line to the ISP. Modems, however, are slow. Faster connections are achieved using DSL, ISDN, cable modems and T1, T2 or T3 lines.

The ISP (Internet Service Provider) Your computer's data connection goes to your ISP. This is a business that links individual users and businesses to the Internet itself. Often an ISP will provide the data connection and the Internet connection.

The Internet

Note that the Internet has many components.

The World Wide Web The WWW or Web is the aggregation of documents that can be accessed by the Internet. These documents include text, graphics, audio, video, photographs and multi-media. The Web is also the aggregation of hypertext servers that can be accessed on the Internet. All Web sites have an "http" prefix attached to the Web site address.

Newsgroups or Discussion Groups These are topic-specific electronic bulletin boards that allow visitors to pose and reply to questions electronically.

E-mail or Electronic Mail These are electronic messages sent from one individual to another. E-mail is one of the most important benefits of the Internet.

How Do I Get an E-mail Account?

If you use a portal such as AOL (American Online) or MSN (Microsoft Network), you can get an account by installing their browser software, connecting to their home page and requesting an account name. If your first choice for account name has already been taken, they will ask you to pick another. You can also get an e-mail account from your ISP.

How Do I Get a Web Address?

The easiest way to get a Web address (domain name) is through an online service where you can select and register a Web address in real time. These services allow you to do an instantaneous search to see if your proposed name has already been taken. Once you find an address you can pay online using a credit card and get a confirmation by e-mail, fax or mail. Note that names that have already been copyrighted cannot be registered by anyone other than the copyright holder.

The most popular international registrar of domain names is Network Solutions (www.networksolutions.com). Domain names are time-limited and must be renewed periodically to keep them active. The cost ranges from US$15 to US$35 per year.

Who Determines How the Internet Works?

There are several important groups that develop standards for and oversee the evolution of the Internet:

World Wide Web Consortium (W3C) The W3C is an industry group supervised by the Laboratory for Computer Science at MIT. The W3C develops standards for the World Wide Web (WWW).

The Internet Society Sponsors the Internet Activities Board (IAB), which works on fundamental architectural and engineering concerns of the Internet.

The Internet Engineering Task Force (IETF) The IETF develops standards on future developments in the Internet's TCP/IP (Transmission Control Protocol / Internet Protocol).

Notes

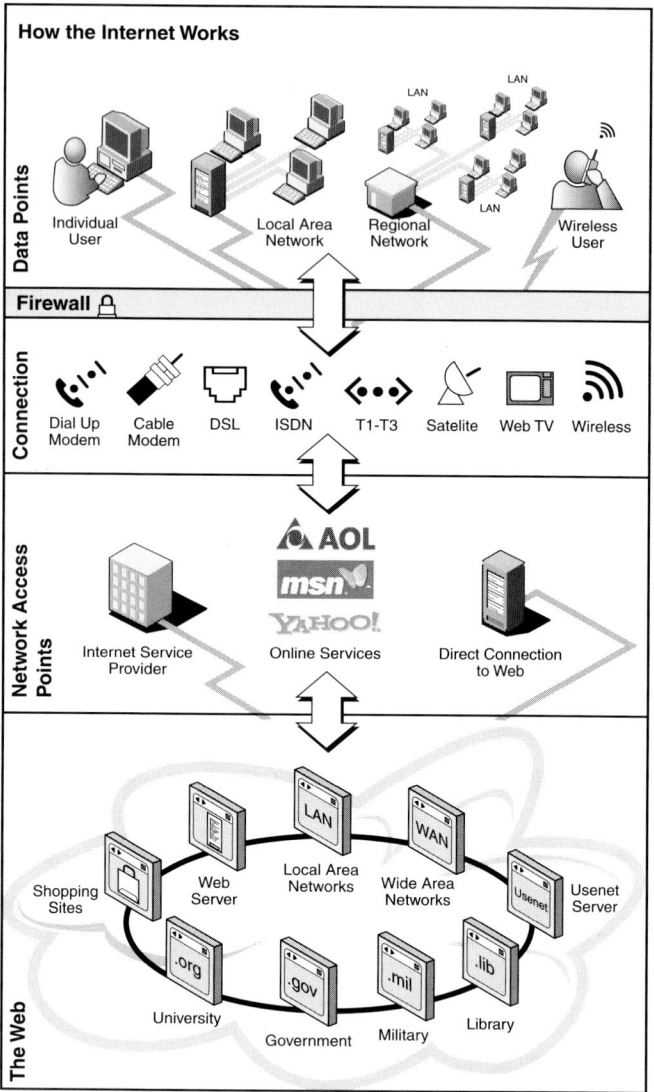

How the Internet Works

Data Points
- Individual User
- Local Area Network
- Regional Network
- LAN
- Wireless User

Firewall 🔒

Connection
- Dial Up Modem
- Cable Modem
- DSL
- ISDN
- T1-T3
- Satelite
- Web TV
- Wireless

Network Access Points
- Internet Service Provider
- Online Services (AOL, msn, YAHOO!)
- Direct Connection to Web

The Web
- Shopping Sites
- Web Server
- LAN — Local Area Networks
- WAN — Wide Area Networks
- Usenet Server
- .org University
- .gov Government
- .mil Military
- .lib Library

Notes

Internet Service Providers (ISPs)

The Basics

ISPs are companies that provide clients with access to the Internet. ISPs provide access in a number of ways.

1. **Dial-up Service** The ISP provides a local access telephone number (a point of presence or PoP) for your computer modem to call over a conventional telephone line. This connects you to the ISP's servers, which are themselves connected to the Internet. You must have a modem in or attached to your computer. Dial-up service is the slowest.

2. **Cable** The ISP, sometimes a local TV cable company, provides Internet access via a TV-type cable wired directly into a home or business. This requires a cable modem attached to your computer. These modems are often provided free of charge. Cable is generally much faster than dial up.

3. **DSL, T1, T3 Lines** The ISP provides a dedicated line to clients. This is the fastest and generally the most costly service available.

Online services such as AOL (America On Line) provide Internet access as part of their service offering.

Travel Issues

There are several issues travelers face regarding Internet access and, therefore, ISPs.

1. A standard ISP connection is a link between a computer in your business or home and the ISP's facility. Essentially, your computer is instructed to call a local number that connects to your ISP. The dialing conventions for making this call are particular to the location of the computer and the ISP. If you change the location of the computer, you may need to change the dialing procedure for connecting to your ISP. See "Re-Configure a Modem" on page 80 to learn how to change these settings.

2. Your established ISP may not have local access numbers at the locations where you are traveling. This means that you may have to place a long-distance call to connect to your ISP. Furthermore, the closest "local" access number may be in another country! While long-distance charges in the U.S. are minimal, in some countries they are substantial, especially when hotels add their surcharges. Furthermore, dial-up connections from some third-world countries to other countries are often of poor quality that may not be able to support data communications at all.

Traveling Using Your Home ISP

If your home ISP offers points of presence (PoP) where you are traveling, you can connect to the Internet using a local access number. Advantages include lower telephone charges and not having to sign up with a new ISP.

Note that ISPs that offer overseas access numbers usually charge a surcharge above and beyond their normal monthly service fee. Check with your provider for both local access numbers and surcharges.

Once you get a local access number you will have to configure your computer to dial that number rather than the at-home access number. See "Re-Configure a Modem" on page 80 for details.

Local ISPs

Most every country has local ISPs. If you plan to stay in a country longer than a few weeks, or if your home ISP doesn't have local access, it may be advantageous to sign up with a local ISP for the duration of your stay. A local connection can be less expensive than dialing internationally to your home ISP.

Finding Local ISPs

There are several ways to find a local ISP. Ask friends and colleagues who have local experience, or consult a website that provides listings. Try "TheList", at **www.thelist.com**, or Topology, **www.topology.org/net/isp_int.html**.

Hotel Internet Service

Many better hotels now provide Internet connections for guests. These are generally in two forms:

1. **Dial-up Service Through In-Room Dataports** (second phone outlets). This requires a modem within or attached to your computer. Note that the phone outlet may be different from your home type and may require a phone plug adapter. See "Phone & Modem Plug Adapters" on page 113 for details.

2. **Ethernet Connection Through an RJ-45 Dataport in the Room or Business Center.** This requires that your computer have an Ethernet port. (Most modern laptops now have Ethernet ports.) This is usually the most economical option for anything beyond a brief occasional connection, especially abroad. Be sure to bring a cable with Ethernet (RJ-45) plugs at both ends.

Questions and Issues

When you make your hotel reservation, ask if they have an Internet connection in the room and if so, if it is dial-up or high-speed service.

If the hotel offers dial-up...

- Make sure you have a telephone cord to plug into the dataport on the room telephone (we recommend a 20-foot retractable type).
- Ask if the phone system data connection is analog or digital. A digital system can ruin your modem.
- Determine in advance what items you'll need to connect successfully. Check "Laptop Travel Kit" on page 95 for a complete list.

If the hotel offers high-speed Internet...

- Can you access the connection in the room?
- Is the high-speed connection through an Ethernet (RJ-45) plug and outlet? Do they have Ethernet cables?
- Is the connection a USB port? How does it work?
- Do you need any other equipment to use their system?
- What are hotel charges for use of the connection?

If the hotel offers wireless service...

- Where can it be accessed in the hotel?
- Does the hotel offer its own Internet service, or does one need an account with another company to use it?
- What are hotel charges to use wireless service?
- Is special hardware required to use their wireless service?

☞ If you plan to use a high-speed data connection, be sure that your laptop has Ethernet capability—either an installed Ethernet capable card and/or a wireless card (or a PCMCIA card to insert in the modem slot).

✐ Refer to "20 Problems & Solutions" on page 123 for more detail on many of these connection issues.

Notes

Global Internet Roaming Services

Internet service providers allow you to connect to the Internet locally, domestically, or internationally, depending on your service. In each country listing, we list selected local service providers as well as the largest U.S. global providers. Global Internet roaming services offer a third alternative. These companies have usually teamed up with local service providers or networks in many countries to provide dial-up and broadband connections at local costs. Sign up from home before departing and have access upon arrival at your destination. Users either receive a software CD or download in order to access the roaming network or they simply receive a password for connection. As always, research to find the most suitable option for your situation. Some roaming services only charge for actual usage, while others have a prepaid subscription rate with usage fees added.

GoRemote Intl. Communications
(formerly GRIC)
1421 McCarthy Blvd.
Milpitas, California 95035 USA
Tel: [1] (408) 955-1920; Fax: [1] (408) 955-1968
European Headquarters
GoRemote Europe Limited
City Point, 1 Ropemaker Street; London EC2Y 9HT
Tel: [44] (20) 7153-1035; Fax: [44] (20) 7153-1135; Web: www.goremote.com

GoRemote Mobile Office provides mobile and remote business people with an easy-to-use software client that allows them to securely connect to the Internet and their corporate networks from thousands of cities in more than 150 countries through dial-up, ISDN, and broadband wireless access points. Featuring state-of-the-art technology, compliant with Wi-Fi standards, the GoRemote network provides more than 9,000 access points outside North America.

Cost: Contact GoRemote

Access Points: 150 countries; 48,000 access points, 16,000 Wi-Fi hotspots, 1300 Hotel Ethernet locations in 10 countries, 500 service providers worldwide

iPass
Headquarters
3800 Bridge Parkway
Redwood Shores, CA 94065, USA
Tel: [1] (650) 232-4100; Fax: [1] (650) 232-4111; Web: www.ipass.com

Mobile professionals receive high-speed broadband access at airports, hotels and convention centers via wired and wireless connections. This VPN-integrated broadband roaming service extends corporate security wherever users travel. Travelers gain the convenience of a single interface for wired and wireless broadband, dial-up, ISDN and PHS.

Access to the iPass network is available through the company's client software, iPassConnect™. A simple point-and-click installation program allows users to easily install, use, and manage the client on laptops, handhelds, and desktops. iPassConnect comes pre-loaded with access points that can be selected by country, state, region, city and/or area code.

Rates: Contact iPass

Access Points: 150 countries

MaGlobe

998 E El Camino Real, Suite #202
Sunnyvale, CA 94087, USA.
Tel: 800-303-2996 (USA only); Tel: [1] (408) 730-9851 (International)
Fax: [1] (408) 730-9853; Web: www.maglobe.com
Branch Office
1355 Mendota Heights Road, Suite #300
Mendota Heights, MN 55120, USA.
Tel: [1] (888) 281-6575 (USA only)
Tel: [1] (651) 365-0500 (International)
MaGlobe is a prepaid service, allowing you to access the Internet while traveling from all worldwide locations with a single Username and Password.
- No set-up / activation fee
- No monthly / annual recurring fees
- Single worldwide username / password
- 800 Broadband Toll Free Internet Access
- One-second billing
- Free 10MB POP E-mail account
- Free AutoDialer for Windows
- Auto-Renewal facility
- Account valid for 1 year from date of activation
Rates: See website
Access Points: 115 countries; 26,500 locations in USA

Net-roamer (GEIR Communications)

GEIR Communications
13499 Biscayne Blvd. #1410
N. Miami, FL 33181 USA
Tel: [1] (305) 944-9333
Toll Free in U.S.: 1-866-944-9333; Web: www.net-roamer.com
Access the Internet through a local call, without changing your e-mail address or ISP through user-friendly software; roaming for laptops, MACs, and PDAs; individual and corporate accounts. Services include dial-up analog connectivity or ISDN. Broadband and wireless access through agreement with iPass.
Rates: US$0.25/minute (US$15/hour)
Access Points: 150 countries, 14,000 access points

WorldWisp (Worldcell)

International Mobile Communications
801 Roeder Road, Suite 800
Silver Spring, Maryland 20910 USA
Toll Free: (888) 967-5323; Tel: [1] (301) 960-0060
Customer Support: [1] (301) 960-0078
Fax: [1] (301) 562-9015;Web: www.worldcell.com
An authorized reseller of the Nokia Data Suite, which allows users to connect a WorldCell handset to a laptop and gain Internet access from virtually anywhere in over 100 countries. Local and international cellular phone rental and satellite phone rental. Worldcell's WorldWisp allows you to connect to the Internet via local dial-up and Wi-Fi access. Point and click and you're online. No more long-distance phone charges.
Cost: See website
Access Points: 140 countries

E-Mail Basics

What Is E-mail?

E-mail or electronic mail is a message sent from computer to computer in digital format over a computer network. The network can be a LAN (Local Area Network) in a small office, a WAN shared by disparate offices of a larger company or, more commonly, the Internet.

In addition to simple text messages, you can also send an e-mail attachment consisting of any digital data file (e.g., word processing file, spreadsheet, illustration, photograph, audio or video).

How Does It Work?

The Software All modern web browsers come with e-mail functionality; however, there are also stand-alone e-mail software packages.

Most people use their Web browser for e-mail. Microsoft's Internet Explorer and Netscape Navigator are the most popular. Services such as AOL (America Online) also offer browser/e-mail software to users.

Your E-mail Address When you first install your browser or e-mail software and connect to your ISP (Internet Service Provider) you will be asked to pick an e-mail address for yourself. Just follow the prompts on screen.

Your address will have three components: 1) your name, 2) a domain name and 3) top-level domain name. For example: your_name@your ISP.com.

If you don't have your own domain name you might use Hotmail, Yahoo or AOL, in which case your e-mail address might be bob_jones@hotmail.com.

✎ Refer to "Glossary" on page 149 for definitions of terms

Sending an E-mail Open your e-mail software and start by inserting the e-mail address of the recipient.

The Message Now write your message in the message window. The software enables you to write, edit and sometimes format your text.

Attachments You may also wish to add an attachment to the message. An attachment can be any file or document on your computer. To add an attachment, find the "Attach" or "Attachment" button on your software. Click once and a dialog box appears asking you to find and select the file or document you want to attach. Once you find the document, click on it once to select it and then click "Attach." This will bring you back to your e-mail software and the name of the attached document will appear in the "Attach" or "Attachment" window. Some e-mail applications allow you to simply "drag and drop" an icon of the file you wish to attach onto the body of the e-mail.

Send When you click "Send", your computer routes your e-mail message (and attachment) through its modem or other data connection port to your ISP, which routes the message over the Internet to the recipient.

Receive When the recipient opens their e-mail software, it notifies them that a message has arrived. For AOL users the famous voice says "You've got mail!" Users are then able to open or download the message and attachment and read it on their computer.

Getting E-mail on the Road

E-mail has become very popular with travelers. There are two sets of issues that must be considered *before* your departure: E-mail Account and Hardware.

E-mail Account The easiest way to check e-mail on the road is through a Web-based system such as hotmail (www.hotmail.com), Yahoo Mail (www.yahoo.com) or juno (www.juno.com). Web-based e-mail is free. Simply sign up on the site, and select an e-mail address and password. You can then get e-mail from any computer in the world that has Internet access. When you leave on a trip you can forward mail from an existing e-mail address to your Web-based (travel) e-mail address. Most travelers who visit Internet cafés use this method. (See "Web-based E-Mail Providers" on page 79 for a list of providers). The disadvantage to free e-mail services is that they restrict the amount of disk space you can take up with your saved messages; these restrictions disappear if you purchase an e-mail subscription.

If you don't want to use Web-based e-mail or an Internet café, you will need to take your computer with you and program it to dial a local (or not as local as you'd like) access number for your ISP and deal with the various issues of mobile connectivity. See "20 Problems & Solutions" on page 123 for details.

Hardware The real issue here is whether you will bring your own laptop or use an Internet café or a friend's or associate's computer. If you bring your own computer you will need electric plug adapter(s), phone plug adapter(s), possibly a transformer and the technical savvy to get a modem connection while on the road. Refer to "20 Problems & Solutions" on page 123 for more details.

Future of E-mail

Single User Interface Common experience is pushing designers to build a single interface to handle e-mail, Usenet News groups, and other groupware forums.

Forms-Based E-mail HTML-formatted e-mail can already include HTML forms. If designed well, such forms can benefit certain types of workflow such as business expense accounts.

E-mail Marketing Laws restricting spam will impact e-mail marketing. E-mail is so cost effective that commercial interests will agree to compromises on spam, most of which will involve the intelligent paring of spam by database-driven, value-added analysis. The commercial e-mail of the future will continue improving accuracy in reaching appropriate parties and avoiding inappropriate ones.

Multimedia E-mail Future increases in bandwidth will enable multimedia e-mail to become inexpensive and (perhaps) ubiquitous in the coming years. Software engineers are doing amazing work with streaming video and audio within e-mails. Multimedia e-mails will rival television commercials in technical quality, with two inherent advantages: they will reside in your mailbox; and you can interact with multimedia e-mail by filling out forms, giving feedback, and, of course, placing orders.

Notes

_____

_____

_____

_____

_____

_____

_____

_____

_____

_____

_____

_____

E-Mail Forwarding

Some travelers may wish to leave their computer at home and forward their e-mail to a friend or associate in the country of travel. We recommend, of course, that you get their permission first. There are two methods for e-mail forwarding: 1) Call your e-mail provider and ask them to forward your e-mail to a new e-mail address for a specific period of time. You may be able to do it yourself online. 2) Reconfigure your e-mail client software to do it for you

☛ **Note**: This is a far less recommended option because <u>anything</u> that goes wrong with the process during your absence, such as power outage, connection termination, etc., will likely preclude any further forwarding of your e-mail until you return.

Reconfigure Client Software to Forward E-mail

The following instructions use Microsoft Outlook / Outlook Express as an example. Other e-mail client software works in a similar manner. In this example you will create a new command or "rule" in your e-mail software.

1. Open Microsoft Outlook, click on the TOOLS drop-down menu and select RULES WIZARD. This opens the Rules Wizard dialog box.
2. Click NEW. A new Rules Wizard dialog box opens.
   This will start the process of creating a new "rule."
3. The dialog box asks "Which Type Of Rule Do You Want To Create?" There are many options in the scroll menu below the question.
   To set up e-mail forwarding, choose
   "Check messages when they arrive" Then Click NEXT.
4. A new window appears and asks the question,
   "Which condition(s) do you want to check?"
   To forward ALL e-mail that is addressed to you, select (click on the box) "where my name is in the To box" Then click NEXT.
5. A new window appears and asks the question,
   "What do you want to do with the message?"
   To forward the e-mail to a specific new e-mail address select:
   "forward it to <u>people or distribution list</u>"
   In the same dialog box in the second white rectangle CLICK
   "<u>people or distribution list</u>"
6. A new dialog box appears called Rule Address. This is where you select the forwarding e-mail address. Note that this address must be listed in the address book which is reproduced on the left side of the dialog box. If necessary click NEW CONTACT, and at a minimum enter the "First" (Name) and forwarding e-mail address" in the "E-Mail Addresses" line. Then click ADD and then click OK.
   Once the e-mail address has been entered, find and double-click the name/address in the address book and the name will appear in the "Specify whom to forward messages to" column. Then click OK.
   This returns you to the previous dialog box. Then click NEXT.
7. This opens a new dialog box window that asks "Add any exceptions", Review the possible exceptions, click those that apply. Then click NEXT.
8. A new window appears that asks "Please specify a name for this rule", Your rule name might be "Forward e-mail to Jack Smith in Buenos Aires", Make sure the "Turn on this rule" box is checked.
   Do not check the "Run this rule now on messages already in "Inbox"" or it will automatically forward everything in your Inbox.
   Then click FINISH.
9. Finally—YOU MUST leave your computer running while you are away.

Suggestion: You may want to make sure your e-mail client is configured to check for e-mail every 5 or 10 minutes. You can configure this by clicking on the TOOLS menu, selecting OPTIONS from the drop-down menu, then select the MAIL DELIVERY tab, and check the box for "Check for new messages every __ minutes." Modify the interval as desired.

Web-Based E-Mail

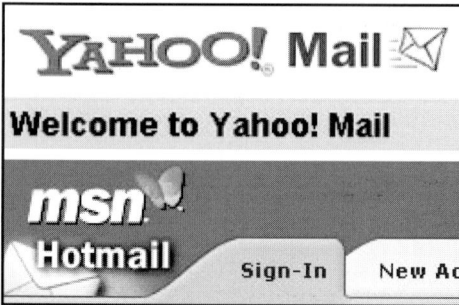

The Problem with Traditional E-mail

One disadvantage of traditional e-mail is that you have to be either sitting at your home or office computer, or you have to program (change settings) on your travel (laptop) computer each time you travel to a new country in order to connect to a local ISP (Internet Service Provider) access telephone number to get or send e-mail. Some niche-market products, such as Blackberry, allow you to have your e-mail sent to your pager-like device; but this is limited to within a single country in all cases.

What Is Web-Based E-mail?

Web-based e-mail is simply access to your e-mail through a service provider's Web site. For example, if you have a "Hotmail" e-mail account (such as <your_name@hotmail.com>), all you need is access to the Internet to access the hotmail.com Web site to access your e-mail. Best yet, there are many Web-based e-mail services that are free to the user.

Getting E-mail on the Road

E-mail has become very popular with travelers. There are two issues that must be considered *before* your departure: E-mail Account and Hardware.

E-mail Account The easiest way to check e-mail on the road is through a Web-based system such as hotmail (www.hotmail.com), Yahoo Mail (www.yahoo.com) or juno (www.juno.com). See "Web-based E-Mail Providers" on page 79 for a list of providers.

Web-based e-mail is free. Simply sign up on the site, and select an e-mail address and password. You can then get e-mail from any computer in the world that has Internet access. When you leave on a trip, you can forward mail from an existing e-mail address to your Web-based (travel) e-mail address. (See "E-Mail Forwarding" on page 77 for details.) Most travelers who visit Internet cafés use this method. The disadvantage to free e-mail services is that they restrict the amount of disk space you can take up with your saved messages; these restrictions disappear if you purchase an e-mail subscription.

If you don't want to use Web-based e-mail or an Internet café, you will need to take your computer with you and program it to dial a local (or not as local as you'd like) access number for your ISP and deal with the various issues of mobile connectivity. See "Re-Configure a Modem" on page 80 and "20 Problems & Solutions" on page 123 for details.

Hardware The issue here is whether you will bring your own laptop or use an Internet café or a friend's or associate's computer. If you bring your own computer, you will need an electric plug adapter(s), phone plug adapter(s), possibly a transformer and the technical savvy to get a modem connection while on the road. Refer to "20 Problems & Solutions" on page 123 for details.

Web-based E-Mail Providers

The following companies/Web sites provide free or low-cost Web-based e-mail or Web-based e-mail forwarding services. Free vs. fee status and rates valid as of 2005.

Note: Many ISPs now offer Web-based e-mail services to their clients. Check your ISP's Web site and/or call for information before you travel.

Alloy Mail
www.alloymail.com   free

Angelfire
www.angelfire.lycos.com   free

ApexMail
www.apexmail.com $24.95/year

Bigfoot
www.bigfoot.com $9.95/year

Cotse*
www.cotse.com $5.95/month

Fastmail*
www.fastmail.fm   free

Flashmail
www.flashmail.com   free

Flairmail
www.flairmail.com   free

Hotmail
www.hotmail.com   free

Hushmail*
www.hushmail.com   free

Juno
www.juno.com   free

Latin Mail
www.latinmail.com   free

Lycos Mail
www.lycos.com   free

Mail.com
www.mail.com   free

MailZone
www.mailzone.com
now Pobox (new accounts with Pobox only)

My Own E-Mail
www.myownemail.com   free

NetAddress
www.netaddress.usa.net $39.95/year; $5.95/monthly

NetForward
www.netforward.com $10.00/year

NTT Verio
http://hosting.verio.co.uk £6.5/month

Pobox
www.pobox.com $20.00/year

ProntoMail
www.prontomail.com $9.99/year

Safemail*
www.safe-mail.net   free

Shanjemail
www.shanjemail.com   fee-based soon

ThatWeb
www.thatweb.com $3.00/month

USA.NET@
www.usa.net $39.95/year

Yahoo! Mail
www.mail.yahoo.com   free

ZapZone
www.zzn.com   free

Zen Search
www.zensearch.com   free

*Indicates that provider allows SSL connection to their servers.

Notes

Re-Configure a Modem

to Dial New ISP Access Number(s)

The Issue

When you connect to the Internet over a telephone line from a dial-up modem, your computer dials a local access number for your ISP (Internet Service Provider). Your computer is set, or configured, to dial this number based on the location of the computer and the location of the access number. The dialing sequence is determined by whether it is a local, long distance, or an international call, and also upon other factors such as the need to dial a special number to get an outside line. If you change the ISP access number and/or location of the computer, you will have to reconfigure the computer/modem to dial the new access number or to dial the original access number from a new location.

First of all, traveling changes the physical location of the computer. Also, you will want to obtain local access telephone numbers in the countries/cities of travel in order to save on telephone charges. Contact your ISP and ask if they have local access numbers in the countries/cities you will be visiting. Most ISPs have this information on their Web sites.

How to Reconfigure Your Modem

The following instructions apply to the Windows 2000 operating system. Instructions for other operating systems and modems will be similar.

1. Go to START –> SETTINGS –> CONTROL PANEL and the Control Panel window will appear. Double-click "Network and dial up connections".
2. In the "Network and Dial-up Connections" window double-click "Make new connection".
3. In the "Network Connection Wizard" window click NEXT.
4. In the "Network Connection Type" window select "Dial up to the Internet," and click NEXT.
5. In the "Internet Connection Wizard" window select "I want to set up my Internet connection manually, or I want to connect through a local area network (LAN)" and click NEXT.
6. In the "Setting up your Internet connection" window select "I connect through a phone line and a modem" and click NEXT.
   (At this time your computer may prompt you to install a modem if it cannot find a modem in the system.)
7. In the "Step 1 of 3: Internet account connection information" window, enter the access number of your ISP and click NEXT.
   Note: Consider where the computer will be when the call is made. For example, if you are configuring your computer in New York prior to departure for a trip to Paris for use in Paris, you will enter the local dialing information for Paris, not the dialing information from New York to Paris.
8. In the "Step 2 of 3: Internet account logon information" window, enter the "User name" and "Password" as you use with your ISP and click NEXT.
9. In the "Step 3 of 3: Configuring your computer" window enter the "Connection name" as provided by your ISP and click NEXT.
10. In the "Set up Your Internet Mail Account" window, select "No" if you already have an account with your ISP and click NEXT.
11. In the "Completing the Internet Connection Wizard" window, select "To connect to the Internet immediately, select this box" and then click FINISH.
12. In the "Dial-up Connection" window click CONNECT.

Exceptions and Other Issues

There are any number of problems that can come up when setting your computer to dial new access numbers from various foreign locations. We recommend that you refer to "20 Problems & Solutions" on page 123 for details.

The Internet Café

What Is an Internet Café?

An Internet café (or cyber café) is a place where people can rent computers with Internet access by the hour and enjoy coffee, tea, soft drinks, light food and the company of other Internet users all at the same time. Many of these establishments provide advice and training for novice users. Internet cafés have become ubiquitous in most metropolitan areas across the world.

Although many people think of Internet cafés as places where backpackers and Bohemian types hang out, the reality is much different. Many business travelers find it easier to go to an Internet café to send and receive e-mail than it is to lug their laptop computer, electric and telephone plug adapters, and transformers with them.

Internet cafés are also very handy for anybody who does not own a computer or who requires special services and equipment that a local café provides.

Users can conveniently use a local Internet café for e-mail, for surfing the net, or for general computer work involving word-processing, spreadsheets, desktop publishing and even advanced specialty software programs.

Many artists such as painters, musicians, and graphics and multi-media designers frequent these cafés; they enjoy the camaraderie, conversation, and collaboration that the cafés make possible.

Note: Users should keep in mind that different countries consider different activities "illegal"; as such, behave online accordingly.

How Do I Find an Internet Café?

Today, thousands of Internet cafés exist in virtually every major and mid-sized city throughout the world.

Global Connect! has listings of many hundreds of Internet cafés throughout the world. The online version of this book at **www.howtoconnect.com** (a subscription site) lists more than 4,500. Once you arrive in a country you can also ask at your hotel for a nearby Internet café.

Other sources of listings on the web are **www.cybercafes.com** and **www.worldofinternetcafes.de**. Both sites include: name, address, e-mail, Web url, facilities, and hours. Note that these are small businesses that start up and go out of business with some regularity. Check ahead whenever possible.

What Services are Provided?

Here is a list of typical services.

Computers and Hardware You will find a wide range of computers from recycled old technology to the most modern and most powerful computers available anywhere. Most cafés have modern equipment to stay competitive. Some locations have powerful computers for people doing multimedia work.

The café should also have CD-ROM, CD-R (recordable CD) or other removable media drives for backing up your work. All should have printers and most will have scanners.

Software Virtually all locations have Web browsers, word-processing, spreadsheet and graphics software. E-mail or call ahead if you have special needs.

Internet Access The café's computer(s) will have a link to the Internet. Most locations have fast access such as DSL, cable or even T1 lines.

VoIP Access Some cafes may now also offer Voice over Internet Protocol (VoIP). See "Voice over Internet Protocol (VoIP)" on page 86 for more details.

Other Services Many cafés offer at least laser printing, color laser printing, fax and photocopying services. Some offer postal services, computer training, Web design, graphic design and Web consulting services.

Food and Drink The range of food and drink varies considerably from café to café. At the low end there may be coffee, soft drinks and sandwiches, and at the upper end, complete meals with beer and wine.

Sockets Alone Some cafés now offer work areas where customers can plug their own laptop computer into a high-speed Internet access port.

Getting Your E-mail

You may well be able to retrieve your e-mail at an Internet café, but it is a separate issue. Refer to "E-Mail Basics" on page 75 for more information. Travelers who use Internet-based e-mail services such as Hotmail or Yahoo Mail, of course, can access e-mail at any time from any place with Internet access.

How Much Does It Cost?

Every café has its own list of services and charges, but they are generally nominal. Expect to pay from a very low US$1 for 20 minutes to a high of US$15, even US$20, per hour for computer and Internet access. Black-and-white laser printing ranges from US$0.25 to US$1 per page. Other services and use of special equipment is extra.

Security Alert!

When you send and receive e-mails from an Internet café, it is likely that a copy remains on the computer's hard drive and will be accessible to either the next user or to someone working at the café. This may be fine if your e-mail simply says that you're having a wonderful time and that the local food is great. However, your e-mail may contain personal or business information that would be either embarrassing or detrimental if it were read by others.

Options:

1) Make certain you log out of your e-mail program before leaving the computer. 2) Don't copy sensitive files onto the computer's hard disk. 3) Don't write anything that you don't want read by third parties. 4) Use encrypted code for sensitive content. 5) Delete e-mail copies from the computer before leaving, and remember to empty the trash. However, even if you intend to take any or all of the above recommended steps, keep in mind that they will not provide protection from the Internet cafe proprietor but only from subsequent users of the terminal you used. Any terminal will likely be configured to prevent users from deleting (or overwriting) files (yours or any other). At the end of each online session, change your user password.

Note: Be aware that in certain "highly structured" countries, all e-mail may be monitored and that making ANY comment about the government, ruling party or person in charge may be punishable by stiff penalties.

Wi-Fi

Wireless Fidelity

What is Wi-Fi?

Wi-Fi is an ultra high frequency (UHF) wireless local area network (WLAN)—a computer network without cables. Wi-Fi allows the user freedom of mobility, just as a cell phone allows callers to roam without wires or cables. Wi-Fi lets computer and PDA users connect to the Internet or to a home or office computer network to share hardware and software resources from a "hotspot" location. In essence, it is a wireless connection that uses radio waves to transmit information at high speeds (at 11 Mbps, almost 200 times faster than a 56K dial-up modem).

How Does It Work?

Existing technologies such as Internet routing protocols, Ethernet's data packets, and the use of a multitude of channels in a frequency band all contribute to how Wi-Fi works.

At a given location, a wired line carrying broadband service connects to a small, inexpensive box called a router/broadband gateway, which distributes wireless broadband Internet access to a wide area called a "hotspot." A card installed in the user's computer (either pre-installed, pre-purchased, or rented at the hotspot) connects to the hotspot via a radio signal, receives the connection, and users are on their way to surfing the Internet or connecting to their office network at any place within the radius of the radio wave (usually about 270 meters). The only wire necessary may be for power.

The Three Wi-Fi Protocols

Referred to as "802.11b" in its initial development, it was soon renamed "Wi-Fi" to be more user friendly. Since then, two other protocols have been developed to deliver faster speeds. The evolution of Wi-Fi is seen in the table below:

Connection speed depends on the distance between the access point and your device, how many active users are using the hotspot, obstructions that may block the signal, the speed of the wired line that connects to the access point, and the software and model of Wi-Fi card that you are using.

Wi-Fi Protocols				
802.11 Protocol	Speed	Band	Advantages	Disadvantages
b	11 Mbps	2.4-GHz band	Compatible with 'g'	Slower speed
a	Up to 54 Mbps usually 6, 12, or 24 Mbps	5- to 6-GHz band	Faster speed than 'b' with less radio frequency interference than 'b' and 'g'	Incompatible with 'b' and 'g'; as such, some manufacturers stopped developing products for this frequency.
g	Up to 54 Mbps	2.4-GHz band	Operates five times faster than the original; compatible with 'b'	More radio frequency interference from other devices operating on the same frequency, such as 2.4 GHz cordless phones

Where Does It Work?

Wi-Fi works anywhere there are other Wi-Fi certified products to equip a connection. Public Wi-Fi stations (hotspots) have sprung up in: corporate environments, airports, hotels, convention centers, libraries, bookstores, coffee shops, Internet cafes, and even parks. Users get connected either through their personal Wi-Fi-equipped laptops (or PDAs) or by using the hotspot's Wi-Fi-equipped desktop computers. However, there is usually a charge for the connection. Each telecommunications provider has lease agreements with its own hotspots and charges for use. At fee-based hotspots, users have to sign up and pay a fee for usage either on the spot or over the Internet with a credit card. Free service may be provided only for certain information. For example, an airport may allow you to check flight information, or, cafés may offer free service to draw customers inside.

Technically, Wi-Fi certified technologies are all compatible, except those manufactured with 802.11a protocol. As such, users need to be aware that different hotspots are designated with different protocols, either 'a', 'b', or 'g'. In other words, you cannot use a 802.11b or 802.11g designated hotspot with an 802.11a device, or vice versa (although most will likely be using 'b' or 'g'). Also, even if you are a monthly Cingular subscriber, for example, you may not have access to Verizon's or T-Mobile's hotspots without paying a separate fee, and vice versa. The different companies and their service sites are working on negotiating roaming agreements to make the process easier.

So, for the basic question, "Can I connect my Wi-Fi equipped laptop at any hotspot?" The answer would be, yes, if your Wi-Fi frequency is compatible with the hotspot's and if you are willing to pay a fee for usage and if you are able and willing to configure the Wi-Fi settings in your computer to those expected by the Wi-Fi provider for which you opted. Check with service providers and their Web sites to see where their hotspots are located.

Keep in mind that hotspots have a limited range of service. As a result, some service providers outfit a high-usage area with plenty of hotspots so that users won't be without a connection when waiting at the gate, going through customs, etc. Philadelphia's Airport, for example, has about 40 hotspots throughout the facility.

Hardware Requirements

In order to get connected, your computer must be configured with a Wi-Fi certified radio. Users basically plug in their card or USB connection, boot up the computer, and begin surfing or file sharing once they've typed in a user ID and password. Computers can use any number of Wi-Fi radio types in order to gain access to the wireless network:

Wi-Fi Hardware Requirements	
Type of Computing Device	**Hardware Applicable**
Desktops	Built-in PCI Card radio (or) Plug-in USB (Universal Serial Bus)
Notebook Laptops	PC card (or) Plug-in USB (Universal Serial Bus built into laptop's motherboard)
Handhelds	CompactFlash
Apple and MacIntosh products	Apple AirPort

Software Requirements

No additional software is necessary for Windows XP and Mac OS X users. Both platforms can detect wireless networks and allow connection without additional software. Other Windows' versions use "drivers" provided by the vendors of the Wi-Fi hardware used. As for PDAs, the iPaz H5550, Toshiba e750, and Palm Tungsten C all already offer Wi-Fi capabilities. Any new notebook on the market from here on out will likely have Wi-Fi capability as well.

Wi-Fi Security

Keep in mind that commercial Wi-Fi providers usually do <u>not</u> enable any encryption on the wireless links. Anything you send or receive may be readily visible to anyone within range; also, unless you have enabled security on your computer, others may be able to roam into any other files present in your computer while you are connected.

The Future of Wi-Fi

Wi-Fi is well on its way to mass usage. Security is being beefed up with a new technology called Wi-Fi Protected Access (WPA), making data more secure. Mobile phones are being developed that can switch between cellular networks and Wi-Fi. However, at the moment, public hotspots are controlled by different providers that charge about US$10 for a 24-hour connection and require the user to fill out an online form with credit card information. Frequent fliers or users of specific airports, hotels, or providers may have access to roaming plans or discount plans, but only at certain hotspots. As such, until cost and somewhat cumbersome formalities of usage at the different hotspots are decreased, the technology may have to wait for the public to adopt it wholeheartedly.

Notes

Voice over Internet Protocol (VoIP)

General Overview

With the advent and growth of the Internet, a new means for making phone calls is becoming increasingly popular. In a nutshell, Voice over Internet Protocol (VoIP) means sending voice signals over Internet lines. Individuals with access to standard Internet connections like dial up, broadband, cable, and satellite can now make voice calls to landline and cell phones, or to other Internet-enabled computers, via the Internet.

How VoIP Works

When making a VoIP phone call, the voice is first converted into digital data (data packets). The data packets then pass through the Internet (just as e-mail does) to the termination point, where they are converted back into recognizable analog sound (voice) for the recipient of the call.

Most users log onto their VoIP account via their PC and make calls directly from one of three applications:

1. **Dialing Application**
   Installed on desktop computer with microphone, speakers (or headset), sound card, and a fast Internet connection. You only pay a monthly ISP fee, most calls are then free.
2. **IP (Internet Protocol) Phone**
   Plugs directly into broadband connection with RJ-45 Ethernet connector. Phones connect directly to router with all appropriate hardware and software for an IP call. Wi-Fi IP phones are on the way.
3. **Internet Phone Adapter**
   (Analog telephone adapter) Plugs into broadband connection and allows regular phone to make VoIP calls by converting analog sound to digital data. Many providers bundle the adapter free with their services. Some adapters come with software for your computer to configure them.

Each of these methods is highly portable. Users can access their accounts from multiple locations as they travel.

VoIP call quality is affected by connection speed. The better the connection, the better the call quality. Generally, a good VoIP service should have call quality equal to or better than a mobile phone, even with slow Internet connection. However, VoIP users in early 2005 reported back that voice quality, installation difficulties, and incoming call problems still left something to be desired. Unreliable access to '911' emergency services via VoIP also necessitates further development work before the public fully latches onto the technology.

Why VoIP Is Less Expensive

With the traditional phone services network, voice is sent as analog sound. This means it is sent as an electrical wave. This wave is variable. Each variable that makes up a part of the wave is important to the overall sound. VoIP calls can be offered less expensively because the digital packets compress to smaller packets of information, requiring minimal bandwidth.

The substantial savings and increased flexibility that VoIP services offer have given the technology a boost of attention. Internet telephony services present discounts of as much as 90% off of traditional phone service rates. VoIP users can access their service in a number of different ways, all of which work particularly well for individuals who travel.

Who Uses VoIP?

Because of the considerable cost savings offered by VoIP, early adopters were large corporations who could save significantly on their communications costs, as well as individual consumers who made frequent international calls. Now, VoIP has become a mainstay for individuals looking to save on their domestic, long-distance and international calling. Currently, many different types of plans can meet individual consumer and business needs.

Internet Cafés

Many Internet cafés outside of the U.S. offer VoIP services to allow their customers to make international calls and take advantage of the low rates that are possible with Internet calls. Cafés will offer either a VoIP calling card, or an Internet-based calling option, or both. You can purchase an account at most cafés for use from any location with Internet access. VoIP calling cards are used in the same way that most prepaid calling cards are offered in the U.S. Some Internet cafés will allow you to make a call using a VoIP service and then pay by the minute at the end of your call. Making international calls while overseas can be much less expensive if you can purchase a VoIP account or use a VoIP service while in an Internet café.

VoIP Providers

Most all major telecommunications and cable TV companies now offer VoIP services. But users can also purchase VoIP calling plans through a variety of VoIP providers listed below.

1TouchTone	www.1touchtone.com
aq	http://aql.com/telecoms/
AT&T	www.callvantage.att.com
BroadVoice	www.broadvoice.com
BroadVox Direct	http://ld.net
Calleveryone	www.calleveryone.com
Earthlink	www.unlimitedvoice.com
Encounta	www.encounta.com
Fastweb	www.fastweb.it
GalaxyVoice	www.glaxyvoice.com
Howdy Corp.	www.hcs.com.sg
iConnectHere	http://ld.net
IP Connection	www.ipconnection.biz
Lingo	www.lingo.com/guWeb
Net2Phone	www.net2phone.com
NuFone	www.nufone.net
Packet 8	http//cognigen.packet8.net
SIP Phone	www.sippone.com
Skype	www.skype.com
SunRocket	www.sunrocket.com
TeliPhone	www.teliphone.ca/en/index.asp
Time Warner	www.twcdigitalphone.com
Unlimitel	www.ulimited.ca
USA Datanet	www.usadatanet.com
Verizon	www22.verizon.com
Voiceglobe	www.voiceglobe.net
Voicepulse	http://voicepulse.com
Vonage	www.vonage.com
VOZ Online	www.vozonline.com
Wazatel	www.wazatel.com

Convergence

Ultimate Connectivity

In the field of wireless communications, convergence refers to the merging of various wireless, computer, satellite, Internet, and software technologies to produce a single device that works anywhere in the world and that acts as a mobile phone PDA (personal digital assistant) with Internet and e-mail access.

The Problem

People who travel with communications technology face a number of problems and limitations: cell phone system standards vary from country to country; cellular coverage outside of urban areas is spotty; traveling with a cell phone, PDA, laptop computer, and accessories is inconvenient and gaining access to the Internet for e-mail while on the road is problematic at best.

The Dream

The dream is simple: one device that does it all.

The Present

The great news is that cell phone and PDA manufacturers are teaming up with satellite system operators to develop the wonder device we had only just begun to think possible. At present, three categories of devices are pushing toward complete convergence: multi-mode/multi-band phones, cellular/PDA combinations, and satellite phones.

Multi-mode/Multi-band Phones are able to send and receive on different frequencies both analog and digital signals. This means that they can work on different cellular service

What's HOT
Late 2005

Blackberry 7250

Nokia 9500

PalmOne Treo 650

systems. For example, some of these phones can operate both in the USA (where there are still some AMPS cellular systems) and in Europe (where GSM is the system standard). Many of these phones are Internet capable. Note that you'll still need to have a service plan that covers both geographic areas.

With the coming dominance of GSM system technology, there are more and more tri-band GSM cellular phones that operate at 900/1800/1900 frequency ranges. This enables the user to make calls on any of the GSM systems around the world.

Cellular/PDA Combinations combine a cell phone with a PDA, an excellent tool for people who organize their lives while on the go. Cellular/PDA combinations usually come with Internet access.

Satellite Phones operate by bouncing their signals off of Earth-orbit satellites. Increasingly, these phones have become multi-mode devices. But since satellite phones are bulky, expensive and have service limitations, many feel that they are the weak link in achieving true convergence. However, once these issues are resolved, satellite phones will be the strong link because of their ability to provide seamless global coverage. Comprehensive global satellite coverage at a reasonable cost remains the challenge.

Current Products

Every few months a manufacturer comes out with the "latest, greatest." Here is a run down of what is available:

Blackberry 7250 33-key keyboard, navigation trackwheel, dual-band CDMA network, e-mail, phone, browser, SMS, Bluetooth, organizer application. Available from Verizon in the U.S. and Bell in Canada.

Nokia 9500 Integrated WLAN 802.11b and EDGE for fast data transfer, full keyboard, office tools, web browser and e-mail with attachments, organizer, PC synchronization, 80MB memory, camera with video, tri-band EGSM 900, GSM 1800/1900 networks, Bluetooth wireless technology, supports Microsoft Office.

PalmOne Treo 650 Combines a mobile phone with e-mail, an organizer, messaging, web access, Bluetooth technology, MP3 player, digital camera, and video. Available as a dual-band digital CDMA phone and as a GSM quad-band world phone

Note: These products do not have satellite capability.

Barriers to Convergence

Some analysts have doubts about the ultimate triumph of convergence. The main barrier, they believe, is simply that the majority of people will never need or be willing to pay more for all the technological extras that come with an "all-in-one" device. That won't stop technology from developing.

Notes

_____

_____

_____

_____

_____

_____

_____

_____

_____

Data Security

How Important Is Your Data?

Information security is one of the hottest topics in the world of connected computers. Business travel often involves the transport of important company documents, which leads to the risk of theft, and the question of data security. Thieves are always on the lookout for laptop computers, cell phones, and other electronic data processing equipment that can be easily sold, allowing sensitive data to be viewed or compromised. Also, unknown persons may view information that is processed through computers at Internet cafés and online. The primary targets for information theft are government and business documents of interest to local intelligence services, business documents of interest to competitors, credit card data, and other identification documents.

While thieves are always finding new ways to crack security, there are a few immediate ways travelers can protect their data from theft and prying eyes.

Data Encryption

Hiding sensitive data through the use of encryption is an important method of protecting data. Encryption helps prevent access to e-mail and important documents. Encryption is useful when passing sensitive materials through the Internet, as well as for Internet users who need to use computers at Internet cafés.

Encryption takes place as data passes through an algorithm where it is converted into encrypted data called ciphertext. The only way to access this data is for the owner of the encryption to possess the "key"; otherwise, deciphering the encryption proves almost impossible.

There are two types of encryption: symmetric and asymmetric. Where symmetric encryption uses one key to access a file, asymmetric encryption uses two keys, a public key and a private key. When users want to send you information, they scramble it using your public key. Upon receiving the information, you are the only one who can de-scramble it with your private key.

Current 256-bit encryption, which is virtually impossible to decrypt, is the industry standard for symmetric encryption. For asymmetric encryption, a key length of at least 4,000 bits is considered to provide equivalent security. Keep in mind, however, that a weak or easy-to-guess encryption key is an obvious weakness. Also, you must make sure that neither the unencrypted version of your file nor the encryption key itself is still residing in your computer.

Note that some countries do not permit foreign nationals to enter the country with encryption software on their computer.

What if Your Laptop Is Stolen?

Anti-theft tracking software and hardware provide a last line of defense against loss of your laptop and data. Unlike encryption that prevents access to sensitive files, theft tracking software and hardware come into play after your laptop is stolen.

Before traveling it is advisable to buy and install the software on your laptop. Once the software is installed, a small, hidden tracking program is activated. This program runs in the background and does not affect your computing. The software communicates using a monitoring point and is usually dependent on the laptop being connected to the Internet where the monitoring program sends the location of the user to a monitoring server or e-mail address. Most software comes with a recovery service that is able to pinpoint the location of the user and is initiated when you report your laptop stolen.

Before purchasing this type of software, consumers should ask whether or not the software includes full recovery costs, and what the software does not protect against. For example, not all tracking software survives the hard drive being reformatted.

Recovering the actual computer, however, is the least of your concerns; the biggest concern is that any sensitive data in it is likely to have been copied. For this reason the data should be encrypted. Do not place any faith in BIOS passwords nor in "password-protected" files, but opt for reliable encryption.

Anti-Theft Tracking Software

Computrace
Web: www.computrace.com

Computrace is one of the most professional tracking programs on the market. The program contacts a monitoring point over the Internet once per day, and more often if reported stolen.

ZTrace
Web: www.ztrace.com

ZTrace is a highly rated tracking program. It contacts its monitoring point, zServer, when a new Internet connection is made. It operates in stealth mode with no trace of its existence on the hard drive.

PC PhoneHome
Web: www.brigadoonsoftware.com

Secretly tracks and locates your missing computer anywhere in the world by sending a stealth e-mail message to a predetermined e-mail address containing your computer's exact location.

Stealth Signal
Web: www.stealthsignal.com

Stealth Signal software offers worldwide monitoring and tracking for Windows and Macintosh computers.

Crossing Borders

Since 9/11, security procedures have been tightened at all airports and other travel points throughout the world. If you travel with a laptop computer, you should expect to be asked to open its travel case and turn it on.

Most airports will ask you to turn on your computer when passing through security. There are some practical ways to ensure that your data will be protected. It is best to leave your computer in suspend mode, thus preventing distractions and theft, and leaving more time to get through the line. Computers are generally stolen from busy x-ray machines, while distracted travelers try to get through security. If your computer has already passed through the x-ray machine, make sure to keep an eye on it and pick it up quickly once it has passed through. If you must go through the metal detector again, ask one of the guards to hold on to your computer while you go do this, but do not leave your laptop on the x-ray machine. Another method to ensure the safety of your computer and data is to remove your hard drive and ask the security staff to hand check it.

Trains and buses pose similar problems. Always keep your laptop, or at least the hard drive, with you. Do not store it in the overhead compartment, as it is too easy for a thief to nab it while you are looking out the window or catching a snooze. For overnight trips, try locking it up with heavier luggage, or lock it to a secure metal bar near you. Pack your laptop in an inconspicuous bag, perhaps even in your suitcase. Computer bags are like beacons to thieves. Again, it is advisable to remove the hard drive and keep it separately so that in the event your laptop is stolen, at least your information is secure. Keep special connection cords packed separately as well.

International Rights of Inspection

While the rights of privacy and freedom of expression are fundamental human rights recognized in all major international and regional agreements and treaties, those rights are being challenged as international threats of terrorism dominate current events. (Under American constitutional law, citizens are protected from search and seizure by the 4th Amendment. This, however, does not protect Americans traveling abroad.)

Data may not only be confiscated by authorities, but also monitored and intercepted at will by government officials. Business travelers who intend to travel into countries where the environment is hostile, or where these rights to privacy might be questioned, should take appropriate measures to ensure that data is secure. Contact the embassy of the country you will be traveling to and ask what procedures to expect—then act accordingly.

Note again that some countries do not permit foreign nationals to enter the country with encryption software on their computer. And some, such as the U.K., can demand that encrypted files be decrypted under penalty of automatic jail sentences.

Internet Cafés

Most travelers prefer to leave their laptop at home and use the resources of the thousands of Internet cafés around the world. The biggest problem with these cafés is that data processed through computers is susceptible to theft. Unless the Internet cafe is capturing keystrokes and/or screen displays, try to connect to your e-mail server or other website through an SSL-encrypted connection. When using a computer at an Internet café, make sure you delete any files you saved on that computer's hard drive, and then make sure to empty the trash. But bear in mind that such "deletions" *(if the Internet cafe's software set-up allows it at all) can be retrieved.

Do not send credit card or other personal information from these computers (or at least make sure the connection is secure), and never check the "save password" box. When you are finished, make sure you log off from any account that you logged into, such as Hotmail or YahooMail.

E-mail Security While Traveling

Generally, generic e-mail that you send or receive through a regular e-mail account is not secured or encrypted to protect the content. Therefore, any personal information you include in an e-mail is at risk of being intercepted by unauthorized individuals.

Do not send sensitive, personal, or financial information unless it is encrypted and is assured encryption from a trusted source. A Web site is said to be encrypted if the web address contains https://, and if a locked padlock symbol appears in the lower right corner of your browser.

☛ **Caution**: Some Web sites and hackers can cause the locked symbol to be displayed even if the connection is not encrypted, giving the user a false sense of security.

Even if the content of your e-mail is encrypted, the connectivity (who is communicating with whom) is not, and neither is the "subject" entry. If it is important to obscure the connectivity, consider the use of "throwaway" e-mail accounts through any of the many online services that offer that, and recommend the same to those with whom you correspond. Some services even allow you to connect to their servers through an SSL connection.

Other Ways to Protect Data While Traveling

- Make sure to back up everything before you leave. Take only the data you will need for business purposes, and leave the rest at home.
- Install a software firewall, and make sure that data is stored in the most secure digital area or applications.
- Utilize encryption and anti-theft software.
- Insure your computer before you travel.
- Prior to departure, install firewall software on your home or business computer to prevent unauthorized individuals or information from entering your computer system.

Laptop Computer Travel Tips

Traveling with a laptop computer has become a necessity for many business and individual travelers. Using travel time productively, the ability to edit presentation materials (such as Power Point presentations), the ability to edit and print important papers (such as contracts) and maintaining access to the Internet and e-mail all argue for taking the laptop along.

The good news is that laptop computers have become smaller, more powerful and less expensive. The bad news is that dealing with security issues, different electrical requirements, different phone plugs, and a host of other connectivity challenges means that you will have to plan travel with a laptop carefully.

Why Are You Taking Your Laptop?

Our first laptop travel tip is: Don't take the laptop!

Ask yourself: Why am I taking my laptop computer with me? Do I really want to lug all that stuff around? How much time do I really expect to be on the computer while traveling? Can I borrow or rent a computer when I arrive? Do I REALLY need it?

Consider the alternatives.

Rent a computer and printer by the hour at your hotel's business center.

Rent a computer at one of the thousands of Internet cafés that have sprung up everywhere, especially in urban areas. Check ahead for locations. Your hotel will be happy to help you find a nearby café.

Need e-mail while on the road? Consider getting a Hotmail or Yahoo Mail account that you can access from any computer with an Internet connection.

Obviously, there are many reasons to take the laptop along. For example, if you are traveling to make an important presentation, it is probably best to have your own computer and not be at the mercy of someone else's poor or malfunctioning equipment. Just be clear about why you're carrying the extra weight.

Laptop Travel Kit

It's always best to have a written list of items to take with you when you travel with your laptop. The first issue to consider is whether you are doing domestic or international travel. If your travel is domestic you are unlikely to need electric and modem adapters, converters, transformers or line testers.

Refer to "Laptop Travel Kit" on page 95 for a list and photo of recommended items to pack for an international trip. That said, be aware that some international trips will require less gear (a trip to a country with the same electrical and modem connections) while others may require more gear (a trip requiring a satellite telephone for Internet connectivity).

Connectivity Issues

There are a multitude of connectivity issues and we've developed a whole section to deal with them. Refer to "20 Problems & Solutions" on page 123 for answers to mobile connectivity issues.

Our experience is that it is wise to read the ENTIRE "20 Problems & Solutions" section PRIOR to departure to understand the nature of connectivity issues.

Predeparture Issues

1. Call your hotel to find out what kind of Internet connection is available in the room (cable, DSL, T1, etc.).
2. Call the hotel and ask what kind of modem plug is required to plug into their system. The standard US (home style) phone plug is an RJ-11, while an Ethernet plug is an RJ-45. You'll need to take the appropriate adapters.
3. Also ask the hotel what kind of electrical voltage and socket types are used in your destination country or countries. You'll need to take along the appropriate adapters and possibly a converter or transformer.

4. Before you leave make sure your primary and secondary laptop computer batteries are fully charged.
5. We recommend that you carry two batteries in addition to the battery in the laptop.
6. Make a backup of important data and application programs so you'll have the data on the computer and on backup disks.
7. Carry backup disks in your suitcase, not your laptop case.
8. Leave backup disks of your data at home or at your office.
9. Take data recovery CDs and floppy disks with you in your CD case. Given that computers do crash, and given the difficulty of starting over while traveling, you may want to prepare and bring with you a fully bootable clone of your hard disk.
10. Set up security devices on your laptop, and password protect your confidential files. But realize that such measures will only deter the casual snoop (e.g., the maid) and not anyone else.
11. Set up a Web-based e-mail account such as yahoo.com or hotmail.com so that you can access your e-mail from anywhere there is an internet connection.
12. Make sure you have a properly working internal dial-up modem in your laptop.
13. For Internet access, make sure you have and carry with you the local access numbers of your home country ISP in your destination country or countries.
14. Make sure you have updated virus protection software installed on your computer.
15. Purchase an inconspicuous laptop carrying case.
16. If your laptop has a CD burner, bring some extra blank CD-Rs so that you can back up important data safely while on the road.
17. Save space by using CD booklets to store your CDs and floppy disks instead of CD jewel cases
18. Make sure you bring the proper phone lines, pocket phone line testers, and phone adapters for your dial-up connection.

While Traveling

1. Due to high airport security, you might be asked to take your laptop out of the case and turn it on.
2. When going through security, be sure to put your laptop through the scanner as you pass the screening device yourself. Many thieves steal laptops from travelers who put their laptop on the conveyor belt and then "get stuck" going through security themselves.
3. Laptop computers are easy to spot and are prime targets of thieves. When standing in an airport or even a hotel lobby, place your computer between your feet to make it more difficult for a thief to snatch and run.
4. When you are on a plane, you can stow your computer under your seat instead of the overhead compartments so that it doesn't get tossed around as much.
5. If you intend to use your laptop in a car, you can save your batteries if you have a car adapter.

Notes

Laptop Travel Kit

The Kit

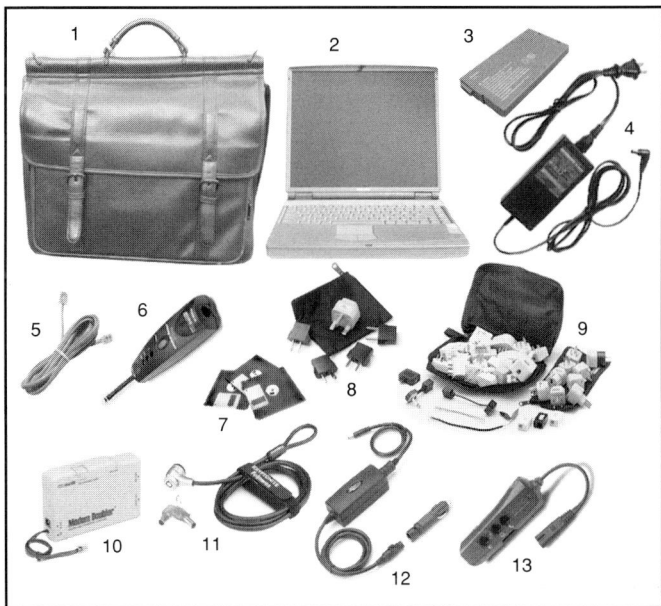

The Checklist

1. Laptop Carrying Case
2. Laptop Computer
3. Extra PDA Batteries (2 recommended)
4. Power Cord / Adapter
5. Phone/Modem Cord
6. Line (telephone) Tester
7. Blank Floppy Disks
8. Electric Plug Adapter Kit
9. Phone / Modem Plug Adapter Kit
10. Digital Line Connector
11. Security Cable
12. Cigarette Lighter Power Cord
13. Surge Protector
14. Portable Mouse (not shown)
15. CD System Recovery Disk (not shown)
16. Emergency Boot Disk (not shown)

Notes

PDA Travel Kit

The Kit

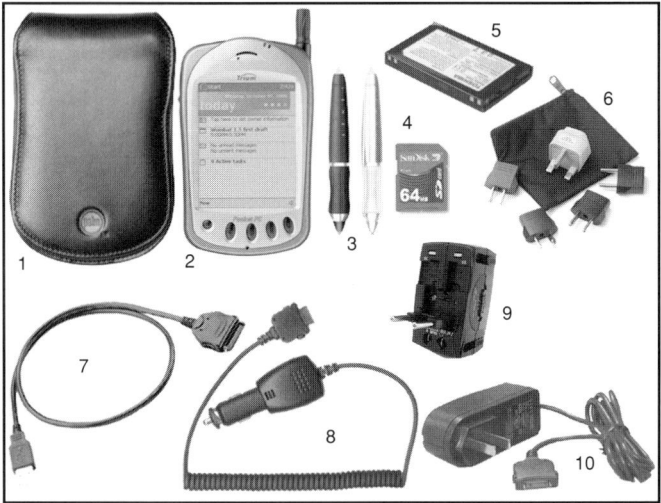

The Checklist

1. PDA Carrying Case   ____
2. PDA (Personal Digital Assistant)   ____
3. Extra Stylus   ____
4. Memory Flash Card (1 - 2 recommended)   ____
5. Extra PDA Batteries (1 - 2 recommended)   ____
6. Electric Plug Adapter Kit   ____
7. Serial Port Adapter   ____
8. Cigarette Lighter Power Cord   ____
9. Electric Power Converter   ____
10. Power Cord / Adapter   ____

Notes

_____

_____

_____

_____

_____

_____

_____

_____

_____

Staying Connected in the Air

Internet and E-mail in the Air

For years air travelers have been using their laptop computers to convert travel time to productive time. Most travelers confine themselves to writing letters and contracts, tweaking Power Point presentations and modifying spreadsheets. In-flight Internet and e-mail connectivity was an unrealized dream.

Recently, however, a growing number of commercial airlines are offering satellite-based Internet and cellular connections for laptops and other electronic devices. First-class and business-class customers have been using these services on a trial basis and have indicated a willingness to pay the premium prices associated with in-flight connectivity. Analysts predict that these services will become widespread in the years to come.

How It Works

Since regulations prohibit the use of cell phones while in flight, some aircraft have been equipped with alternatives which involve the use of aircraft-provided equipment that allow passengers to connect to terrestrial telephones for a usually steep fee. Some of these use terrestrial stations, and some of the newer ones bounce signals off of geosynchronous-orbiting satellites to transmit and receive data. Special satellite antennas are installed on the airplane.

Boeing's Connexion service, the most ambitious plan to date, uses an "always on" real-time technology, while some less expensive competing services store data on a server and connect periodically for transmission and for uploading updates.

A "connected" seat has a power outlet and a network plug underneath it, which means that the traveler doesn't have to bring along and use up extra batteries.

Requirements, Services and Cost

For a laptop to work with in-flight connection systems, it must have a network card installed (an Ethernet card, for example). Boeing's Connexion also includes a local wireless network, so that laptops with wireless network cards installed can use the services without even plugging in. Other less-ambitious in-flight connection systems that do not employ wireless networks require the network card and a physical plug-in to the network. Most recent laptops come automatically with this hardware configuration, although travelers should double check their laptops just to be sure.

Lufthansa was the first to commit to installing Boeing's high-speed, real-time Internet Connexion service on its long-haul aircraft from Frankfurt and Munich, charging US$29.95 per flight or $9.95 per 30 minutes and $0.25 cents a minute thereafter. Scandinavian Airline System (SAS), All Nippon Airways, China Airlines, Asiana, and Korean Air already offer service or have signed deals. British Airways abandoned plans for the wireless Internet after initial tests, citing high costs. The airline will wait and see how Lufthansa's Wi-Fi works out before making any further plans. The three major U.S. airlines pulled out early on in development also citing high costs.

Connexion aims at a 1Mbps transmission, which is comparable to cable modems. Multiple simultaneous transmissions among more than one traveler can slow this rate down, but to never less than 56Kbps.

In addition to laptop computers, other devices that can take advantage of Connexion are PDAs, including the new "convergent" PDA/cellular phone devices. Connexion is committed to implementing "voice-over" Internet technology, which will enable the connected traveler to make telephone calls via laptops or PDA/cellular phones. Currently, in-flight phone calls are something of a luxury since they require the use of pricey on-board phones.

An example of a less-expensive, simpler system is JetConnect, which is produced by the U.S. telecommunications provider Verizon. JetConnect employs on-board computer servers that "batch" transmissions and receptions, and then at regular intervals of 15 minutes connect to the Internet (through the same geosynchronous satellite technology) to transmit and receive the accumulated, batched information. JetConnect is anticipating adding high-speed e-mail. Surfing the Internet is not an option. JetConnect costs the connected traveler US$5.99 per flight segment, take-off to landing.

Notes

Electricity—Basic Issues

Volts, Watts, Hz—What's It All About?

We use it daily, but electricity is a complete mystery to most people. For many, it is enough simply to plug their computer into an electric wall outlet. For travelers, however, it is a different story; different electrical requirements, electric outlets and plugs require some basic knowledge of electricity.

There is hope. In the next few pages we'll introduce issues and offer solutions for common electrical problems for those who travel with electronic devices.

The Issues:

Issue	Explanation / Solution
Different countries use different electrical systems.	This usually means that your device or appliance from home operates on a different voltage than that of the country you are visiting. For example, if you plug a 110-volt device into a 220-volt outlet you are likely to destroy the device. You may need a "converter" or "transformer". The good news is that most modern laptop computers have converters built in to the power cord. See "Electric Appliance Labels" on page 100. See "Electric & Phone Plug Chart" on page 106. See "Electric Converters/Transformers" on page 103.
Different ountries use different electric outlets and plugs.	This means that your device uses one style of plug and the electric outlet in another country may use another. You will need to get a "plug adapter" for each country or outlet type you will encounter. See "Electric Plug Adapters" on page 102. See "Electric & Phone Plug Chart" on page 106. See "World Electric Plugs" on page 105.
Do I need a transformer or a converter?	Transformers and converters take the electrical voltage from the electric outlet and alter it to be compatible with your electric appliances and devices. The words transformer and converter seem easily interchangeable, but there is a difference. Converters are for electronic devices such as computers, while transformers apply to electric appliances such as hair dryers and items with electric motors. See "Electric Converters/Transformers" on page 103.

ampere (amp or amps) Electrical flow or current.

electric appliance Any device that uses electricity to power a heating element or a motor. Low wattage (to 50 watts): radios, portable CD players, shavers and contact lens cleaners. High wattage (50 to 1600 watts): garment steamers, hair dryers, irons, heating pads and curling irons.

electronic device Any device that uses electricity to power electronic (computer) circuits: computers, printers, scanners and battery chargers.

electricity The flow of electrons in a circuit (a wire or electric motor or computing device).

frequency The number of repetitions of a complete electrical waveform in a stated period. Usually stated as cycles per second.

volt A measurement of electrical force. All electric appliances and machines are designed to operate using a particular voltage. In the U.S., the most common voltage is 110-120 volts. In the rest of the world it is 220 volts.

wattage A measurement of electrical power or the amount of "work" produced by electricity. 1 watt = 1/746 horsepower. Volts x Amps = Watts.

Electric Appliance Labels

Reading Product Identification (Electric) Panels

All electronic devices and appliances have a manufacturer's product identification label—on the bottom or back of most devices and anywhere it fits on others. This label identifies the manufacturer, product, model and other information, but specifically the device's electrical requirements as follows:

AC or DC — Means the product uses DC (Direct Current) or AC (Alternating Current). In DC the electric current flows in one direction in the circuit. In AC the electric current flows both ways, alternating at a specific frequency. In the U.S., most household appliances operate using AC at 60 cycles per second, whereas most laptop computers operate on DC.

Voltage or V — Specifies the voltage (measurement of electric force) that the device was designed to use. U.S. appliances are designed for 110 volts, but most of the rest of the world operates on 220 volts.

Amps or A or Amperage — A unit of measure of the rate of flow of electric current. This is usually not an issue unless the device "draws" a great deal of current, and the electric outlet is on an already overloaded circuit.

Htz or Hz — Abbreviation for Hertz. A unit of measurement of the frequency at which an alternating current (AC) flows back and forth expressed in cycles per second. For example, 60 Hz means 60 cycles per second.

Wattage — A unit of measurement of the rate at which electric energy is used, especially by light- or heat-dissipating devices such as light bulbs or hair dryers.

An Appliance Label

With electric appliances both voltage and wattage are important. Appliances use either a heating element and/or a fan.

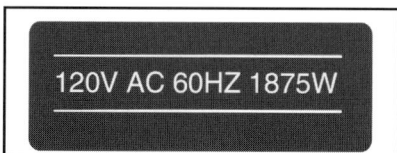

120V AC 60HZ 1875W

Above This hair dryer label states "120 v (volts) 1875W" (watts). It is high-wattage because it is above 50 watts. If you travel to a country that uses 220 volts you will need a transformer to use this appliance.
See "Electric Converters/Transformers" on page 103.

An Electronic Device Label

With electronic devices the voltage is the most important factor. Any device that uses any electronic circuitry is an electronic device.

SONY® MODEL PCG-9211
NOTEBOOK COMPUTER
DC 19.5V ⎓ 2.15A
Tested To Comply
With FCC Standards
FOR HOME OR OFFICE USE
The Windows logo is a registered
LISTED
13C1
I.T.E.

Above This label is for a laptop computer and states "DC 19.5V" (volts). This means that the device requires input voltage of 19.5V and the parellel symbol after it means DC voltage at 2.15 amps, much less than either the U.S.

standard (110V) or the European standard (220V). It is important to notice what is stated after INPUT. In the next image, the 100-240V designated after "INPUT:" means that you do not need a transformer at all; only an adapter to ensure that the plug will fit into the wall receptacle.

Above This is the label for the AC Adapter of the same laptop computer. Note the label states "Input 110v-240v~, Output 19.5v." This means that it will take any voltage input from 110 to 220 volts AC and convert it to the laptop computer's required 19.5 volts DC. With this adapter you will be able to use the above computer with any power input of 110 to 220 volts. (The transformer is built into the power cord itself.)

For more details see "Electric Converters/Transformers" on page 103.

Notes

Electric Plug Adapters

Different Electric Outlets and Plugs

Different countries of the world may use different electric plugs and outlets from those you're used to at home. If the electric plug of your electronic device or appliance does not match the outlet in the country you are visiting, you will not be able to "plug in" and get electric power.

☛ In addition to being able to plug in and get electric power, your electronic device or appliance must be compatible with the electric power that comes out of the outlet. See **Caution!** below for more information.

Electric Plug Adapters

An electric plug adapter is simply a plastic and metal part that fits between the electric plug of your device and an electric outlet.

Each plug adapter is unique to the plug being inserted into the adapter and the electric outlet into which the adapter is plugged. Therefore, if you travel to different countries you may need a different electric adapter for each. Also, if you travel with electronic devices or appliances that have different types of plugs (for example, you purchased one device in the USA and another in Germany), you will need different electric plug adapters for each device, for each country you visit.

Sample 1:

Electric plug adapter (center) for U.S. plug to fit German outlet.

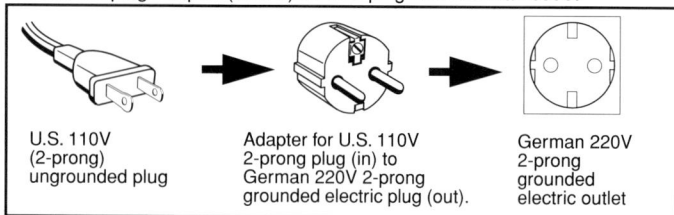

U.S. 110V
(2-prong)
ungrounded plug

Adapter for U.S. 110V
2-prong plug (in) to
German 220V 2-prong
grounded electric plug (out).

German 220V
2-prong
grounded
electric outlet

Sample 2:

Electric plug adapter (center) for German plug to fit U.S. outlet.

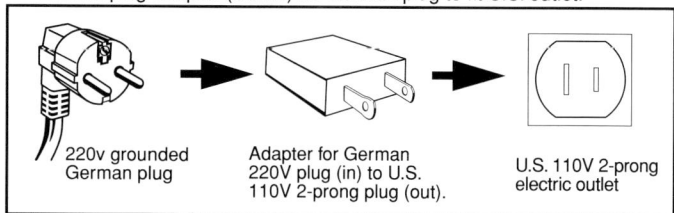

220v grounded
German plug

Adapter for German
220V plug (in) to U.S.
110V 2-prong plug (out).

U.S. 110V 2-prong
electric outlet

✐ To learn which type of electric plug is used in different countries, see "Electric & Phone Plug Chart" on page 106.

✐ For electric plug adapters see "Mobile Connectivity Suppliers" on page 114.

Caution!

☛ An electric plug adapter solves the problem of getting electric current from a wall outlet to a device or appliance. It DOES NOT solve the problem of making sure that the electric voltage level coming from the wall outlet is appropriate for the electronic device or appliance you are using.

✐ Refer to "Electricity—Basic Issues" on page 99, "Electric Appliance Labels" on page 100, and "Electric Converters/Transformers" on page 103 for more information.

Electric Converters/ Transformers

Converters and Transformers

Converters and transformers take the electrical current from an electric outlet and alter it to be compatible with your electric appliance or electronic device. The words converter and transformer seem easily interchangeable, but there are very important differences.

Converters

Use a converter for *electric appliances* ONLY. Electric appliances have either heating elements or motors and include:

- hair dryers
- curling irons
- fans

Converters are used only for short periods of time (generally 10 minutes to 2 hours).

Different converters are available for appliances with 50-1857 watt ratings.

☛ Be ABSOLUTELY certain that the wattage capability of the converter you purchase is sufficient to run your appliance. Look for the wattage rating on the label.

Cautions

It is important to use the correct product, or you could risk blowing out your appliance or device. Converters are available in a range of 50 to 1875 watts, and transformers in ranges of 50, 100, 200 watts or more.

☛ The wattage of your appliance or device must fall within the transformer's range; to be on the safe side, give it a 10- to 20-watt buffer.

☛ When in doubt, use a transformer with the appropriate wattage.

✍ See "Mobile Connectivity Suppliers" on page 114.

Notes

_____

_____

_____

_____

_____

_____

_____

_____

_____

_____

Transformers

Use a transformer for BOTH *electronic devices* and *electric appliances*. Electronic devices include:

- computers, printers
- battery chargers
- any device with a computer chip

Travelers requiring continuous use of an appliance or device (continuous use for more than one or two hours) should bring along a "continuous use transformer," regardless of whether the appliance could be used with a converter.

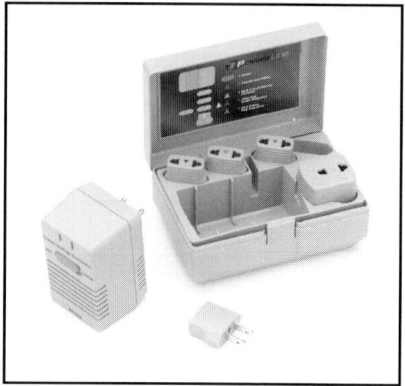

A "2 to 1 *step-down* transformer" will take 220-volt electricity and enable you to run a 110-volt device or appliance.

A "1 to 2 *step-up* transformer" will take 110-volt electricity and enable you to run a 220-volt device or appliance.

☛ Transformers with a capacity greater than 200 watts are heavy. Also, they are often more expensive than purchasing a second appliance with the needed voltage or a dual-voltage appliance.

☛ A high-low combination with a low-wattage transformer and a converter for 50-1875 watts in a single unit is also available. It is imperative to ensure that this device's switch is set to the "low" setting (i.e., the transformer setting) when powering electronic devices, <u>BEFORE</u> anything is connected.

☛ Most laptop computers come with a transformer on the power cord that adapts 110~220 volt current to the required current of the computer.

☛ Most transformers sold are what are known as "autotransformers" (i.e., have a single coil tapped at mid-range) rather than true transformers. The rare problem with this is that the device one connects to it may end up "hot", i.e., at a voltage other than zero, which can give an electrical shock to a user who touches the device's metal chassis and a water pipe at the same time, or may burn the device when that device is connected to a different electrical circuit such as the telephone line (in the case of fax machines and some laptops).

Notes

World Electric Plugs

Notes:

1. There is no universally accepted name for each plug type.
2. Each adapter supplier will have its own stock number depending upon the individual country's electric outlet AND the plug type of the electric appliance.
3. Tthe required adapter will be determined by both the plug type of your electronic device AND the socket type of the country you are visiting.
4. See "Mobile Connectivity Suppliers" on page 114 for electric plug adapters.

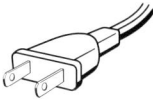

EP-1
Standard USA

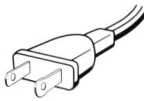

EP-2
Polarized EP-1

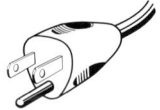

EP-3
Grounded EP-1

EP-4

EP-5

EP-6
Grounded

EP-7
Grounded

EP-8
Grounded

EP-9
Grounded

EP-10
Grounded

EP-11
Grounded

EP-12
Grounded

EP-13

EP-14
Grounded

EP-15
Grounded

Electric & Phone Plug Chart

For plug images, see "World Electric Plugs" on page 105, and "Phone & Modem Plugs" on page 112.

COUNTRY	ELECTRICITY			TELEPHONE
	Volts	Frequency	Plug	Plug
Afghanistan	220V	50 Hz	EP-4,10	TP-D
Albania	220V	50 Hz	EP-4	TP-D, S
Algeria	127/220V	50 Hz	EP-4,6	TP-C
Angola	220V	50 Hz	EP-4	TP-A, L
Argentina	220V	50 Hz	EP-4, 15	TP-A, X
Armenia	220V	50 Hz	EP-4, 6	TP-D
Australia	220/240V	50 Hz	EP-15	TP-A, I
Austria	220V	50 Hz	EP-4, 6	TP-Y
Azerbaijan	220V	50 Hz	EP-4, 6, 10	TP-D
Bahamas	120V	60 Hz	EP-1, 3	TP-A
Bahrain	220V	50 Hz	EP-10, 12	TP-A,B
Bangladesh	220V	50 Hz	EP-1, 4, 6, 10, 12	TP-A, H
Belarus	220V	50 Hz	EP-4, 6	TP-A, B
Belgium	220V	50 Hz	EP-4, 6, 7	TP-A, Z
Belize	110/220V	60/50 Hz	EP-1, 3, 12	TP-A, B
Benin	220V	50 Hz	EP-4, 5	TP-C, E
Bermuda	120V	60 Hz	EP-1, 3, 12, 15	TP-A
Bolivia	110/220V	60/50 Hz	EP-1, 3, 4, 6	TP-A
Bosnia & Herzogovina	220V	50 Hz	EP-4, 6, 10, 12	TP-G
Botswana	230V	50 Hz	EP-10, 12	TP-B
Brazil	110/220V	60/50 Hz	EP-1, 3, 4, 10, 12	TP-A, C, AA
Brunei	220V	50 Hz	EP-11, 12	TP-A, B
Bulgaria	220V	50 Hz	EP-4, 6	TP-D
Burkina Faso	220V	50 Hz	EP-4, 7	TP-C
Burundi	220V	50 Hz	EP-4, 6, 7	TP-E
Cambodia	120V	50 Hz	EP-4	TP-A
Cameroon	220V	50 Hz	EP-4, 6, 7, 10, 12	TP-C, E
Canada	110V	60 Hz	EP-1, 3	TP-A

| COUNTRY | ELECTRICITY | | | TELEPHONE |
	Volts	Frequency	Plug	Plug
Central African Republic	220V	50 Hz	EP-4, 7	TP-E
Chad	220V	50 Hz	EP-6, 7, 10	TP-C
Chechnya	220V	50 Hz	EP-5	TP-D
Chile	220V	50 Hz	EP-4, 9	TP-A
China	220V	50 Hz	EP-4, 12, 15	TP-A, BB
Colombia	110/220V	60/50 Hz	EP-1, 3, 4	TP- A, CC
Comoros	220V	50 Hz	EP-4, 7	TP-C
Congo, Rep. of (Brazzaville)	230V	50 Hz	EP-4, 5	TP-C
Congo, Dem. Rep. of (Kinshasa)	220V	50 Hz	EP-4, 7	TP-C
Costa Rica	110/120V	60 Hz	EP-1, 3	TP-A
Côte d'Ivoire	220V	50 Hz	EP-4, 7	TP-A, E
Croatia	220V	50 Hz	EP-4, 6	TP-G
Cuba	110V	60 Hz	EP-1, 3, 6	TP-A, D
Cyprus	220V	50 Hz	EP-4, 7, 12	TP-A, B
Czech Republic	220V	50 Hz	EP-4, 6, 7	TP-DD
Denmark	230V	50 Hz	EP-4, 6, 7	TP-L
Diego Garcia	110/220V	60 Hz	EP-1, 3	TP-A
Djibouti	220V	50 Hz	EP-4, 6, 7	TP-C
Dominican Republic	110V	60 Hz	EP-1, 3	TP-A
Ecuador	120V	60 Hz	EP-1, 3, 4, 10	TP-A
Egypt	220V	50 Hz	EP-4	TP-A, C
El Salvador	115/120V	60 Hz	EP-1, 3	TP-A
Equatorial Guinea	220V	50 Hz	EP-4, 7	TP-A, C
Eritrea	220V	50 Hz	EP-4, 11	TP-A,B
Estonia	220V	50 Hz	EP-6	TP-D
Ethiopia	220V	50 Hz	EP-4, 6, 7, 9, 10, 11	TP-J
Fiji	240V	50 Hz	EP-15	TP-I
Finland	220V	50 Hz	EP-4, 6, 7	TP-NN

| COUNTRY | ELECTRICITY | | | TELEPHONE |
	Volts	Frequency	Plug	Plug
France	220/230V	50 Hz	EP-4, 6, 7	TP-C
French Guiana	230V	50 Hz	EP-4, 6, 7	TP-A, C
Gabon	220V	50 Hz	EP-4, 6	TP-C
Gambia, The	220V	50 Hz	EP-12	TP-C
Georgia	220V	50 Hz	EP-4, 6	TP-D
Germany	220V	50 Hz	EP-4, 6	TP-A, N, O, P, Q, R
Ghana	220V	50 Hz	EP-4, 10, 11, 12	TP-A, B, H
Greece	220V	50 Hz	EP-4, 6, 7, 10	TP-A, S
Guam	110V	60 Hz	EP-1, 3	TP-A
Guatemala	120V	60 Hz	EP-1, 3, 12, 15	TP-A
Guinea	220V	50 Hz	EP-4, 6	TP-A
Guinea Bissau	220V	50 Hz	EP-4, 6	TP-A
Guyana	110V	50/60 Hz	EP-1, 3, 4, 10, 12	TP-A
Haiti	110V	60 Hz	EP-1, 3	TP-A
Honduras	110V	60 Hz	EP-1, 3	TP-A
Hong Kong	220V	50 Hz	EP-10, 12	TP-A, B
Hungary	220V	50 Hz	EP-4, 6	TP-Y, EE
Iceland	220V	50 Hz	EP-4, 6	TP-A
India	220V	50 Hz	EP-4, 6, 10, 12	TP-H
Indonesia	110/220V	60/50 Hz	EP-4, 6, 7	TP-A, I
Iran	220V	50 Hz	EP-4, 6	TP-A, EE, F
Iraq	220V	50 Hz	EP-4, 10, 12	TP-A, F
Ireland	220V	50 Hz	EP-6, 12	TP-A, B
Israel	220V	50 Hz	EP-3, 4, 6	TP-FF, GG
Italy	220V	50 Hz	EP-4, 6, 9	TP-J, K
Jamaica	110V	50 Hz	EP-1, 3	TP-A
Japan	100V	60 Hz	EP-1, 3	TP-A, HH
Jordan	220V	50 Hz	EP-4, 6, 7, 10, 12	TP-A, B, U
Kazahkstan	220V	50 Hz	EP-4, 6	TP-D
Kenya	220/240V	50 Hz	EP-10, 12	TP-A, B, H
Korea, North	110V	60/50 Hz	EP-1, 3, 4, 6, 10, 12, 15	TP-V

| COUNTRY | ELECTRICITY | | | TELEPHONE |
	Volts	Frequency	Plug	Plug
Korea, South	110/220V	60/50 Hz	EP-1, 4, 6, 10, 12, 15	TTP-A, V
Kosovo	220V	50 Hz	EP-4, 6	TP-G
Kuwait	220V	50 Hz	EP-4, 11, 12	TP-A, B, U
Kyrgyzstan	220V	50 Hz	EP-4, 6	TP-D
Laos	220V	50 Hz	EP-1, 3, 4, 6, 7	TP-A
Latvia	220V	50 Hz	EP-4, 6	TP-D
Lebanon	110/220V	50 Hz	EP-1, 3, 4, 7, 10	TP-A
Lesotho	220V	50 Hz	EP-4, 11	TP-A, B, E
Liberia	120V	60 Hz	EP-1, 3, 12	TP-A
Libya	127/230V	50 Hz	EP-9, 10	TP-A
Lithuania	220V	50 Hz	EP-4, 6	TP-D
Luxembourg	220V	50 Hz	EP-4, 6	TP-N
Macau	220V	50 Hz	EP-4, 10	TP-A, L
Macedonia	220V	50 Hz	EP-4, 6	TP-A, B
Madagascar	127/220V	50 Hz	EP-4, 6, 7, 10, 15	TP-C
Malawi	220/240V	50 Hz	EP-12	TP-A, B, L
Malaysia	220V	50 Hz	EP-12	TP-A, B
Maldives	220V	50 Hz	EP-1, 4, 7, 9, 10, 11, 12	TP-A
Mali	220V	50 Hz	EP-4, 6, 7	TP-C
Malta	220V	50 Hz	EP-12	TP-B
Mauritania	220V	50 Hz	EP-4	TP-C
Mauritius	220V	50 Hz	EP-4, 12	TP-A, C
Mexico	120V	60 Hz	EP-1, 3	TP-A
Moldova	220V	50 Hz	EP-4	TP-A, D
Mongolia	220V	50 Hz	EP-4, 10	TP-D
Morocco	220V	50 Hz	EP-4, 6, 7, 10	TP-C
Mozambique	220V	50 Hz	EP-4, 6	TP-A, L
Myanmar	220V	50 Hz	EP-4, 6, 10, 12	TP-A, B
Namibia	220V	50 Hz	EP-4, 10, 11	TP-E
Nepal	220V	50 Hz	EP-4, 10	TP-A
Netherlands	220V	50 Hz	EP-4, 6	TP-M

| COUNTRY | ELECTRICITY | | | TELEPHONE |
	Volts	Frequency	Plug	Plug
New Zealand	220V	50 Hz	EP-15	TP-A, B
Nicaragua	120V	60 Hz	EP-1, 3	TP-A
Niger	220V	50 Hz	EP-1, 3, 4, 6, 7, 10	TP-C
Nigeria	220V	50 Hz	EP-10, 12	TP-B, H
Norway	220V	50 Hz	EP-4, 6	TP-NN
Oman	240V	50 Hz	EP-4, 10, 12	TP-A, B
Pakistan	230V	50 Hz	EP-4, 10, 11	TP-A
Panama	120V	60 Hz	EP-1, 3, 15	TP-A
Papua New Guinea	240V	50 Hz	EP-15	TP-A, I
Paraguay	220V	50 Hz	EP-4	TP-A
Peru	110/220V	60/50 Hz	EP-1, 3, 4	TP-A
Philippines	110/220V	60/50 Hz	EP-1, 4, 6, 7, 15	TP-A
Poland	220V	50 Hz	EP-4, 6, 7	TP-D
Portugal	220V	50 Hz	EP-4, 6, 7, 10	TP-A, L
Qatar	240V	50 Hz	EP-10, 12	TP-A, B
Romania	220V	50 Hz	EP-4, 6	TP-A
Russia	220V	50 Hz	EP-4, 6	TP-D
Rwanda	220V	50 Hz	EP-4, 7	TP-C
Saudi Arabia	110/220V	60/50 Hz	EP-1, 3, 4	TP-A, B, C, T, U, II
Senegal	127/220V	60/50 Hz	EP-4, 7, 10	TP-C, E
Serbia & Montenegro	220V	50 Hz	EP-4, 6	TP-G
Sierra Leone	220V	50 Hz	EP-10, 12	TP-H
Singapore	220V	50 Hz	EP-4, 10, 12	TP-A, B
Slovak Republic	220V	50 Hz	EP-4, 6, 7	TP-D, G
Slovenia	220V	50 Hz	EP-4, 6	TP-G
Solomon Islands	220V	50 Hz	EP-15	TP-I
Somalia	110/220V	50 Hz	EP-4, 15	TP-C
South Africa	220V	50 Hz	EP-11, 12	TP-E
Spain	220V	50 Hz	EP-4, 6, 7	TP-A

| COUNTRY | ELECTRICITY | | | TELEPHONE |
	Volts	Frequency	Plug	Plug
Sri Lanka	220V	50 Hz	EP-4, 10	TP-A, B
Sudan	220V	50 Hz	EP-4, 6, 10, 12	TP-J
Suriname	120V	60 Hz	EP-4, 6	TP-M
Swaziland	220/230V	50 Hz	EP-4, 11	TP-B
Sweden	220V	50 Hz	EP-4, 6	TP-JJ
Switzerland	220V	50 Hz	EP-4, 6, 7	TP-LL, MM
Syria	220V	50 Hz	EP-4, 6, 7, 9, 10	TP-A, F
Taiwan	110V	60 Hz	EP-1, 3, 15	TP-A
Tajikistan	220V	50 Hz	EP-4, 6	TP-D
Tanzania	220V	50 Hz	EP-4, 10, 12	TP-B
Thailand	220V	50 Hz	EP-1, 3, 4	TP-A
Togo	220V	50 Hz	EP-4	TP-C
Trinidad & Tobago	110/220V	60/50 Hz	EP-1, 3, 10, 12	TP-A
Tunisia	127/220V	50 Hz	EP-4, 6, 7, 9	TP-C
Turkey	220V	50 Hz	EP-4, 6	TP-A, F
Turkmenistan	220V	50 Hz	EP-4, 6	TP-A, D, F
Uganda	220V	50 Hz	EP-10, 12	TP-A
Ukraine	220V	50 Hz	EP-4, 6	TP-D
United Arab Emirates	220V	50 Hz	EP-6, 10, 11, 12	TP-A
United Kingdom	220V	50 Hz	EP-12	TP-A, B
United States	110V	60 Hz	EP-1, 2, 3	TP-A
Uruguay	220V	50 Hz	EP-4, 9, 15	TP-A
Uzbekistan	220V	50 Hz	EP-4, 6	TP-D
Venezuela	120V	60 Hz	EP-1, 3	TP-A, CC
Vietnam	110/220V	60/50 Hz	EP-1, 3, 4, 6	TP-A
Western Sahara	220V	50 Hz	EP-4, 6, 7, 10	TP-C
Yemen	220/230V	50 Hz	EP-1, 4, 10, 12	TP-A
Zambia	220V	50 Hz	EP-4, 10, 12	TP-B
Zimbabwe	220V	50 Hz	EP-10, 12	TP-B, E

Phone & Modem Plugs

Notes:
1. With a few exceptions there is no standard name for each plug type.
2. Each adapter supplier will have its own stock number depending upon the individual country's outlet AND the plug type used.
3. See "Mobile Connectivity Suppliers" on page 114 for adapters.

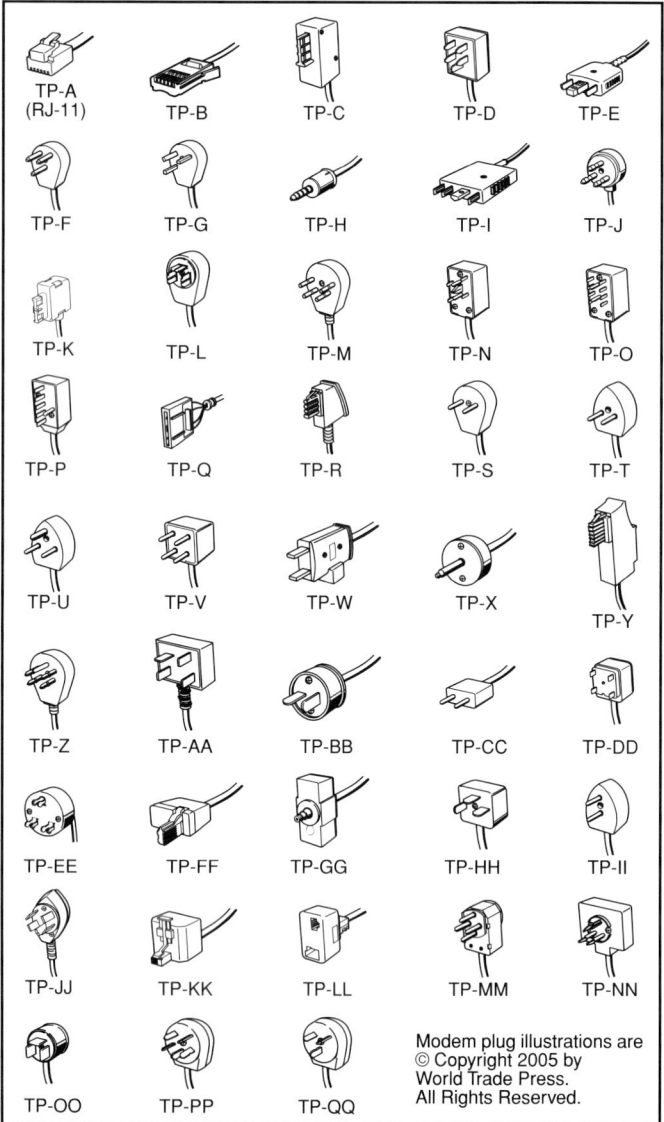

TP-A
(RJ-11)

TP-B

TP-C

TP-D

TP-E

TP-F

TP-G

TP-H

TP-I

TP-J

TP-K

TP-L

TP-M

TP-N

TP-O

TP-P

TP-Q

TP-R

TP-S

TP-T

TP-U

TP-V

TP-W

TP-X

TP-Y

TP-Z

TP-AA

TP-BB

TP-CC

TP-DD

TP-EE

TP-FF

TP-GG

TP-HH

TP-II

TP-JJ

TP-KK

TP-LL

TP-MM

TP-NN

TP-OO

TP-PP

TP-QQ

Modem plug illustrations are
© Copyright 2005 by
World Trade Press.
All Rights Reserved.

Phone & Modem Plug Adapters

Modem Outlets and Plugs

Different countries of the world may use different phone and modem plugs and outlets from those you're used to at home. If the phone plug of your modem does not match the phone outlet in the country you are visiting you will not be able to "plug in" and get a phone/data/Internet connection.

In addition to being able to plug in and get a data connection, you will need to know if the connection is analog or digital. Computer modems use analog signals. Many hotels and offices, however, use digital phone systems, which are unable to process analog signals, in which case one needs to use a work-around, which amounts to connecting one's laptop (through a suitable adapter) through the connection between the telephone instrument and the handset, which is almost always "analog". See "Are the Phone Lines Digital or Analog?" on page 130 for more information.

Modem Plug Adapters

A phone/modem adapter is simply a plastic and metal part that fits between the modem plug and the telephone or data outlet.

Each plug adapter is unique and specific to the plug being inserted into the adapter and the phone/data outlet into which the adapter is plugged. Therefore, if you travel to different countries you may need a different adapter for each. Also, if by chance you travel with several different modem-type devices (such as a laptop and a PDA) that have different types of modem plugs (for example, you purchased one device in the USA and another in Germany), you will need different adapters for each device, for each country you visit.

Sample

The illustration below shows a phone plug adapter (center) for U.S. RJ-11 plug to fit Turkish phone outlet.

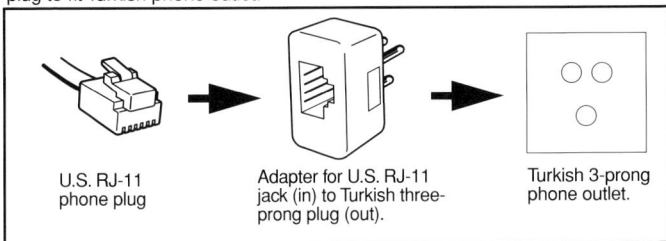

| U.S. RJ-11 phone plug | Adapter for U.S. RJ-11 jack (in) to Turkish three-prong plug (out). | Turkish 3-prong phone outlet. |

✏ To learn which type of phone/modem plug is used in different countries, see "Electric & Phone Plug Chart" on page 106 and "Phone & Modem Plugs" on page 112.

✏ For phone/modem plug adapters see "Mobile Connectivity Suppliers" on page 114.

Caution!

☛ A phone/modem plug adapter solves the problem of making a connection between your phone/modem plug and the phone/data terminal.
It DOES NOT solve the problem of making sure that the data signal is a compatible analog or digital connection.

☛ Refer to "Are the Phone Lines Digital or Analog?" on page 130 for more information. **Warning**: do not plug your modem into a digital line, it can ruin your modem and/or computer. When in doubt, plug into fax lines, which are always analog and safe to plug into.

Mobile Connectivity Suppliers

USA

Absolute4.com
832 Parkside Blvd.
Claymont, DE 19703 USA
Toll Free USA (877)520-5748
Tel: [1] (302) 793-0166
Fax: [1] (302) 793-1449
E-mail: info@absolute4.com
Web: www.absolute4.com

Austin House Inc. (USA)
P.O. Box 665
Amherst, NY 14226 USA
E-mail: custserv@austinhouse.com

iGo
9393 Gateway Drive
Reno Nevada 89511 USA
Tel: [1] (888) 205-0093 (toll-free in USA)
Tel: [1] (775) 746-6140
Web: www.igo.com
E-mail: helpdesk@igo.com

Konexx
5550 Oberlin Drive
San Diego, CA 92121 USA
Tel: [1] (800) 275-6354 (toll-free in USA)
Tel: [1] (858) 622-1400
E-mail: support@konexx.com
Web: www.konexx.com (in U.S.)

Laptop Travel
P.O. Box 46106
Plymouth, MN 55446 USA
Tel: [1] (888) 527-8728 (toll-free in USA)
Tel: [1] (763) 404-9496
E-mail: mail@laptoptravel.com
products@laptoptravel.com
Web: www.laptoptravel.com

Magellan's
110 W. Sola Street
Santa Barbara, CA 93101 USA
Tel: [1] (800) 962-4943 (toll-free in USA)
Tel: [1] (805) 568-5400
E-mail: orders@magellans.com
customerservice@magellans.com
Web: www.magellans.com/

Mobile Planet
9175 Deering Ave.
Chatsworth CA 91311 USA
Tel: [1] (800) 675-2638 (toll-free in USA)
E-mail: sales@mobileplanet.com
customerservice@mobileplanet.com
Web: www.mplanet.com

Mobilex, Inc.
4225 Executive Square #1170
La Jolla, California 92037 USA
E-mail: lclee@mobilex.com
Tel: [1] (858) 558-5445
Fax: [1] (858) 558-5450
Web: www.mobilex.com

Port.com
1211 North Miller Street
Anaheim, CA 92806 USA
Tel: [1] (800) 950-5122 (toll-free in USA)
Tel: [1] (714) 765-5555
Web: www.port.com

ProMax Wireless Products
10415 Westpark Dr., Suite B
Houston, TX 77042, U.S.A.
Tel: [1] 713-974-2277
Toll Free (USA only): (888) 314-7926
Fax: [1] (713) 974-1102
E-mail: pda@promaxwireless.com
Web: www.promax

R.S. Communications
2701 Stanley Gault Parkway #1
Louisville, KY 40223 USA
Tel: [1] (502) 254-0443
Web: www.rs-comm.com

Travel Oasis
PO Box 2191
St. Peters, MO 63376-0040 USA
Tel: [1] (877) 894-1960 (toll-free in USA)
Fax: [1] (636) 922-1979

Teleadapt (USA)
1762 Technology Drive
Suite 223, San Jose, CA 95110 USA
Tel: [1] (877) 835-3232 (toll-free in USA)
Tel: [1] (408) 350-1440
E-mail: info@us.teleadapt.com
Web: www.teleadapt.com

Walkabout Travel Gear (USA)
P.O. Box 1115,
Moab, UT 84532 USA
Tel: [1] (800) 852-7085 (toll-free in USA)
Tel: [1] (888) 722-0567 (North America)
Tel: [1] (435) 259-4974 (international)
Tel: [44] (207) 681-2776 (Europe/UK)

Notes

Canada

Austin House Inc. (Canada)
2880 Portland Drive
Oakville, Ontario L6H 5W8 Canada
Tel: [1] (905) 829-0111
Web: www.austinhouse.com

Globus Wireless
1955 Moss Court
Kelowna, British Columbia V1Y 9L3
Tel: [1] (604) 860-3130
Web: www.globuswireless.com

Mexico

iGo Mexico
Tel: [52] (55) 170-0234
Web: www.igo.com.mx
E-mail: info@igo.com.mx

Palm Mexico
Web: www.palm.com/mx/index.html
E-mail: palm_soporte@next.com.ar

Nextel de México
Blvd. Manuel Avila Camacho No. 36

Col. Lomas de Chapultepec
Del. Miguel Hidalgo; C.P. 11000,
México, DF
Tel: [52] (55) 1018-3300
Tel: (01-800) 200-9300 (toll-free within
Mexico)
Web: www.nextel.com.mx
E-mail: info.ventas@nextel.com.mx

United Kingdom

A Phone Accessories
PO Box 23374
Rotherhithe
London SE16
Tel: [44] (161) 798 5470
E-mail: info@a-website.net
Web: www.a-phone-accessories-website.co.uk

Avrmobiles
Unit 128, Park House
15-19 Greenhill Crescent,
Watford Business Park
Watford WD18 8PH
United Kingdom
Tel: [44] (1923) 819900
Web: www.avrmobiles.co.uk

Beyond 2K Cellular
Villandro House
Freetown way
Kingston upon Hull
East Yorkshire
HU2 8JL
England (UK)
Tel: [44] (0)870 756 3700
Web: www.b3k.net

Cellmania UK
Viewpoint
Basing View
Basingstoke
Hants, RG21 4RG
United Kingdom
Tel: [44] (1256) 347800
Fax: [44] (1256) 347801
Web: www.cellmania.com

Connexions
52 Chichester Road
Tilehurst, Reading, RG30 4XB
United Kingdom
Tel: [44] (0)79 73 753212
Fax: [44] (0)118 9623009
E-mail connexions@freeuk.com
Web: www.connexions.f2s.com

IBM (United Kingdom) Ltd.
PO Box 41; North Harbour
Portsmouth, Hampshire PO6 3AU,
England U.K.
Tel: 0-8705-426-426 (from the U.K.)

Phatphones
Unit 2, Bounds Green Station,
Bounds Green Road,
London N11 2EU, UK
Tel: [44] (0)20 8888 1000
Fax: [44] (0)20 8922 3470
Web: www.phatphones.com

TeleAdapt Ltd. (UK)
The Technology Park
Colindeep Lane
London NW9 6TA United Kingdom
Tel: [44] (20) 8233-3000
E-mail: worldtraveller@teleadapt.com
Web: info@uk.teleadapt.com

UKPhoneShop
Freedom House
18 Parliament Street
UPHolland, Wigan
Lancashire WN8 0LN
United Kingdom
Tel: [44] (0) 870 330 8860
Fax: [44] (0) 1695 633213

Notes

Europe Excluding Germany

AGS srl
Via Enrico Fermi, 56
24035 Curno, Bergamo
Italy
Tel: [39] 035 615659
Fax: [39] 035 613421
E-mail:info@agsricambicellulari.com
Web: www.agsricambicellulari.com

AsianComm OU
Peterburi tee 47
11415 Tallinn, Estonia
Tel: [372] 621 0433
Fax: [372] 621 0435
E-mail: info@assiancomm.ee
Web: www.asiancomm.ee

Benefon
P.O. Box 84; Meriniitynkatu 11
FIN-24101 Salo, Finland
Tel: [358] 2-77 400
Fax: [358] 2-733 2633
Web: www.benefon.com/
products.index.htm

Breskvar Trade
Breskvar Darjan s.p.
ïhlava 5/a
9244 Sv. Jurij Ob ≥cavnici
Slovenia
Tel: [386] (2) 56 89 141
Fax: [386] (2) 56 89 140
Web: www.breskvar-trade.si

CCM Europe N.V.
Waaslandlaan 8A5

Define Cellular Accessories
Zahrievskaya St. 23A
Saint Petersburg 191123
Russian Federation
Tel: [7] (812) 3273830
E-mail: info@define.ru
Web: www.define.ru

Euroset
Department 1-076, 1 floor
Barclay St. 8, Trade Center "Gorbushkin
Dvor"; Moscow, Russia
Tel: [7] (095) 737-4676
Web: www.evroset.ru

IBM France
BP 51; F-45802 - Saint Jean de Braye;
France
Tel: [33] 238-557-777 or 0801-835-426

MCA Accessories
48bis, quai de Jemmapes
75010 Paris, France
Tel: [33] 01 42 01 08 88

Fax: [33] 01 42 01 09 99
E-mail: info@manhattan-cellular.com
Web: www.manhattan-cellular.com

Moscow store:
proezd Airport 11
Moscow 125167
Russian Federation
Tel: [7] (095) 726-5655
E-mail: msk@define.ru

Motorola A/S
Motorola Kundeservice
Dept. A101; Box 0553
1532 Copenhagen V; Denmark
Tel: [45] 4348 8005
Fax: [45] 8001 0755
E-mail: info.dk@motorola.com
Web: http://danmark.motorola.com

Oskar
Cesk∆ Mobil a.s.
Vinohradská 167
Prague 10
Czech Republic
Tel: (800) 777-777 (toll-free in the Czech
Republic)
Web: www.oskarmobil.cz

PhoneTrade.com
GBL Global Business Link AB
P. O. Box 14370
SE-400 20 Göteborg; Sweden
Tel: [46] (31) 733 33 44
Mobile: [46] (709) 46 19 29
E-mail: info@phonetrade.com
Web: www.phonetrade.com

ProMax Wireless Products
Ana Bajo
C/Leitariegos, 10; 28761 Tres Cantos
Madrid, Spain
Tel: [34] 91-804-9505
Fax: [34] 91-804-9506
ana@promaxwireless.com

Telefonomania
Via Vittorio Veneto, 135/A
Monopoli (Bari), Italy
Tel: [39] 080 9306068
Fax: [39] 080 4170175
E-mail: info@telefonomania.biz
Web: www.telefonomania.biz
9160 Lokeren - Belgium
Tel: (32) (9) 340-60-80
E-mail: ccm@ccm.be
Web: www.ccm.be

Notes

_____

_____

_____

_____

Germany

Amazon.de
www.amazon.de
Bosch
Bestell-Hotline
Zubehör und Ersatzteile GSM-Produkte
Tel: [49] (0)180 500 23-41
Fax: [49] (0)180 500 23-55
Web: www.bosch.de/start/de/products/
mobil/index.htm
IBM Deutschland (Germany)
Am Fichtenberg 1,
71083 Herrenberg, Germany
Tel: [49] 0180-331-3233

Mobilcom Ag
Hollerstrasse 126
D-24782 Buedelsdorf
Germany
Tel: [49] 43 31 69 11 73/74
Mobstar GmbH
Bayerwaldstr. 25
81737 München; Germany
Tel: [49] 89 6734690
Fax: [49] 89 63897643
E-mail: info@mobstar.com
Web: www.mobstar.com

India

Cellmania India Private Limited
A30 Kailash Colony
New Delhi, 110048
India
Tel: [91] (11)6430538
Fax: [91] (11)6486738
Web: www.cellmania.com

IBM India, Ltd.
Golden Towers
Airport Road, Bangalore
India - 560 017
Tel: [91] (80) 526 7117
Fax: [91] (80) 527 7991

Japan

IBM Japan Ltd
2-12 Roppongi, 3 Chome
Minato Ku, Tokyo 106-8711 Japan
Tel: [81] 3 3586 1111
Tel: 0120 04 1992 (toll-free in Japan)
Web: www.ibm.co.jp
Oi Electric CO., Ltd.
3-16 Kikuna 7-Chome
Kohoku-Ku Yokohama222-0011
Japan
Tel: [81] 45 4331361, 45 4018044

Okinawa Cellular Telephone Company
14-1, Kumoji, 2-Chome
Naha Okinawa Pref. 900-8540
Japan
Tel: [81] (98) 8691001
Fax: [81] (98) 8692643
Web: www.tccsecure.com

Others in Asia

Aletech International Ltd
Unit 4
20/F Cheung Fund Ind. Bldg.
23-29 Pak Tin Par Street
Tsuen Wan. NT
Hong Kong
Tel: [852] 2493 8828
Fax: [852] 2493 8618
Aifa Technology Group
230 Bai Der 2nd Rd.
Fong Shan City, Kaohsiung Hsien
Taiwan R.O.C.
Tel: [886] (7) 7434157
Fax: [886] (7) 7412190
Web: www.aifa.com.tw
Cellware International Co., Ltd
7F, NO.67, Sung Po Street
Pan Chiao City, Taipei
Taiwan, R.O.C.
Tel: [886] (2) 82511060
Fax: [886] (2) 22544783
E-mail: cellware@ms35.hinet.net
Web: cellware2000.com.tw

Formosa Electronic Industries Inc.
P.O.Box 10255
Hsin Tien City, Taipei Hsien
Taiwan, R. O. C.
Tel: [886] 2-2218-8888
Tax: [886] 2-2218-8889
E-mail: sales@feii.com.tw
Website: http://www.feii.com
IBM China/Hong Kong Limited
10/F PCCW Tower
Taikoo Place, 979 King's Road
Quarry Bay, Hong Kong
Tel: [852] 2825-6222
E-mail: ibmhkg@vnet.ibm.com
IBM Singapore Pte. Ltd.
80 Anson Road ; IBM Towers
Singapore 079907
Tel: [65] 320 1000
Tel: (toll free in Singapore) (1 800) 320
1000; Fax: [65] 224 5260
E-mail: direct@sg.ibm.comd

IBM South Africa
Private Bag x9907
70 Rivonia Road
Sandton 2146, South Africa
Tel: [27] (11) 302 9111
Tel: (toll free in South Africa) (0-800) 130-130
Fax: [27] (11) 302 6161
E-mail: ibm4you@za.ibm.com

Jingyou Communication Tech Co.
3/F, Bldg 1
Songshan Industrial Zone
Shatou Village, Shajing Town,
Bao'an District
Shenzhen City, Guangdong Province,
China 518104
Tel: [86] (755) 27211778,27203930
Fax: [86] (755) 27203980
Web Site: http://www.szjingyou.com

ProMax Wireless Products
Roger Liao
3B-04, No. 5, Sec. 5
Hsin Yi Rd., Taipei, Taiwan
Tel: [886] 2-27233930
Fax: [886] 2-27229429
E-mail: roger@promaxwireless.com

TeleAdapt (HK) Ltd.
Unit 711-712 7/F Peninsula Tower,
538 Castle Peak Road
Kowloon, Hong Kong
Tel: [852] 2780 9020
Fax: [852] 2780 9019

Australia

Access Communications Pty Ltd
PO box 231 Northbridge
NSW 1560 Australia
33-35 Alleyne Street Chatswood
NSW 2067 Australia
Tel: [61] (2) 9417-5311
Fax: [61] (2) 9417-6976
E-mail: sales@accesscomms.com.au

Cellmania Australia
2 Lambert Court, Endeavour Hills
Victoria - 3802
Australia
Tel: [61] (4) 12525049
Fax: [61] (3) 97060591
Web: www.cellmania.com

Cellphone Technology
PO Box 11235
Papamoa, TAURANGA
NEW ZEALAND
Tel: [64] (27) 488-5776
E-mail: Sales@4phones.co.nz
Web: www.4phones.co.nz

IBM (Australia)
55 Coonara Avenue; P.O. Box 400
West Pennant Hills
New South Wales; Australia 2120
Tel: [61] (2) 9354 4000
Web: www.ibm.com

South Africa

IBM South Africa
Private Bag x9907
70 Rivonia Road
Sandton 2146, South Africa
Tel: [27] (11) 302 9111
Tel: (toll free in South Africa) (0-800) 130-130
Fax: [27] (11) 302 6161
E-mail: ibm4you@za.ibm.com

Shawcell Telecommunications Ltd
136 Cross Street
Kroonstad 9499
South Africa
Tel: [27] 56 2125255, 56 2122751
Web: www.shawcell.com

Notes

Important Web Sites

Communications

www.howtoconnect.com

Global Connect! On-line. The world's most comprehensive telecommunications, cell communications and mobile connectivity resource. All the data of the library edition of Global Connect! (1184 pages) plus six additional data sets for each of 175 countries.

Country Data

www.bestcountryreports.com

Automated-download country reports for 175 countries. Categories include: Business Culture, Communications, Travel, Security, Economics and Trade, Trade Documentation and others. Cost from $6.50 to $45.00 per report.

www.cia.gov/cia/publications/factbook

The "CIA Factbook." Offers concise country-by-country information on geography, people, government, economy, communications, transportation, military and transnational Issues. Information is in the public domain.

Courier Services

www.ups.com

United Parcel Service. World's largest express carrier and package deliverer, and a leading provider of specialized transportation, logistics, capital, and e-commerce services.

www.dhl.com

DHL Courier. A global leader in the international air express industry, operating in more than 220 countries.

www.fedex.com

Federal Express. Global corporation offering an array of transportation, e-commerce and supply chain solutions.

www.couriernetwork.com

Ark International Couriers, Ltd. Overnight delivery to most major cities in Europe, Canada, Latin America, and the Caribbean. Delivery within 48 hours to the Far East and other major international cities.

www.skypak.com

SkyPak. European-based package delivery services to more than 200 countries. Delivers materials up to 30 kilograms in weight. Online service lets you monitor the transit and delivery status of your shipment minute-by-minute 24/7.

Currency Rates, Exchange, and Purchase

www.oanda.com

Comprehensive global currency information site. Lets users purchase foreign currencies; contains currency tools for travelers and business people.

Embassy and Consulate Information

www2.tagish.co.uk/Links/embassy1b.nsf

Comprehensive links to web sites and other information on embassies of various countries around the world.

www.embassyworld.com

Comprehensive list of contacts for intl. diplomatic offices.

http://travel.state.gov

Links to U.S. embassies and consulates worldwide.

Health

www.cdc.gov

Center for Disease Control. The U.S. Government's site for disease control and global immunization information.

www.cyberpharmacy.com

Ask questions, research health topics, search prescription prices and request prescriptions online. Has menu-driven library of educational materials.

Hotels

www.hotelbook.com

Lets users research and reserve hotel rooms in countries across the globe.

www.hotelchoice.com

Rates and availability for rooms at major hotel chains.

www.all-hotels.com

Online travel, lodgings, and reservations for 77,000 luxury, chain and discount hotels worldwide.

Internet Cafés

The following are typical sites providing updated information on Internet cafés around the globe:

www.cybercafes.com

www.worldofinternetcafes.de

Internet Service Providers

www.thelist.com

A resource for people looking for an Internet service provider in countries around the world.

www.nsrc.org/networkstatus.html

Major emphasis on ISPS in Asia, Africa, Latin America and the Caribbean, the Middle East, and Oceania.

International Trade

www.worldtradeREF.com

Authoritative, comprehensive data for international trade, documentation and logistics. Used extensively by traders, freight forwarders and logistics professionals. Subscription based.

www.fita.org

Federation of International Trade Associations (FITA). A source for B2B leads, news, events and links to 5,000 international trade-related Web sites.

www.trade-circuit.com

International business opportunities. Includes free Import-Export leads, Agents wanted and sought, Joint venture and venture capital. Useful trade-tools.

Language

http://babelfish.altavista.com

AltaVista Babel Fish translates short passages of text or entire Web sites among 19 pairs of languages.

www.freetranslation.com

FreeTranslation lets users obtain free translations of both text and web pages. Covers English, Spanish, French, German, Italian, Dutch, Portuguese and Norwegian.

Maps

www.mapquest.com

Mapquest. Major source of city and country maps.

www.maps.com
Provides digital maps, travel guides and gear, an online map store, driving directions and an address locator.

News
The major portals such as Yahoo, AOL, Lycos, etc. all contain menu-driven and searchable news databases.

http://news.google.com
Excellent up-to-the-minute global coverage.

www.bbc.co.uk
British Broadcasting Corporation's news from around the world. Country profiles.

www.cnn.com
Leverages CNN's global team of almost 4,000 news professionals. Has searchable archives and multi-media news presentations.

www.csmonitor.com
The Christian Science Monitor is an independent, international daily newspaper published Monday through Friday. It stands out because it does not rely primarily on wire services, like AP and Reuters, for its international coverage. It has writers based in 11 countries.

www.reuters.com
The world's largest international multimedia news agency that supplies news – text, graphics, video and pictures – to media organizations and websites around the world.

Restaurants

www.zagat.com
Zagat Survey is a long-established guide to the top restaurants in the USA and in major cities around the globe.

www.restaurants.com
Continuously updated site with information on restaurants in countries around the world.

Search Engines / Portals

www.google.com
Leading search engine on the Internet. Currently searches well over three billion web pages. Also contains group message boards and menu-driven content.

www.yahoo.com
Most visited Internet portal in the world. Offers Google-based searches as well as a huge menu-driven database of information and news. Independent Yahoo sites are maintained for the major world languages, countries, and regions.

www.aol.com
AOL: Major portal famous for providing non-technical users with easy Internet access, e-mail, and messaging. Has a big menu-driven database of news and information.

Time

www.globaltimeclock.com
The time in 231 countries of the world.

www.dataandtime.com
Comprehensive site that gives the time and date for every country in the world.

www.worldtime.com
Features an interactive world atlas, information on local times and a public holiday database for countries around the world.

Travel (General)

www.globalroadwarrior.com
Global Road Warrior. 175-country business travel, business communications and business culture resource. "The resource for businesspeople, not back-packers."

www.expedia.com
Microsoft's online travel service features a fare finder, discounts, destination information, vacation packages, maps, global guidebook search, accommodations, transportation and travel merchandise.

www.btonline.com
Business Traveler Online. The online version of the monthly magazine for business travelers featuring country profiles, plus hotel, airline, and special perks information.

www.lonelyplanet.com
A travel guide to many destinations around the world, aimed at the budget and adventure traveler, but with many helpful tips and insights useful to business persons.

www.travelocity.com
Travelocity. Online travel service with air, hotel, and car reservations, maps, weather, currency converter, electronic ticketing, destination guide, consolidator fares, travel agency locator and travel headlines.

www.wtgonline.com
World Travel Guide. Extensive travel site with key information for most countries, including business protocol and contact information. Has separate guides relating to cities, weather, airports, events. Has an online bookstore.

Travel Documents

www.traveldocs.com
Country-by-country listing of required travel documents and customs information. Works closely with the U.S. Passport agency and embassies of many countries worldwide.

Travel Advisories & Security

http://travel.state.gov
Consular Information Sheets. U.S. Department of State country information sheets apprising travelers of current events and dangers affecting international travel safety.

www.fco.gov.uk/travel
The UK government's official travel-advisories site.

www.voyage.gc.ca
Canadian government's official travel advisories page.

www.dfat.gov.au/consular/advice
Australian government's official travel advisories page.

Weather

www.wunderground.com
This comprehensive global site provides concise weather data for cities and towns in most countries of the world

www.w3.weather.com
Provides weather reports by region and by topic (driving, travel, ski, etc.). Covers USA, Brazil, France, Germany, Latin America, and the UK.

www.earthcam.com
Live webcam images of outdoor conditions worldwide.

20 Problems & Solutions

20 Problems & Solutions for Mobile Connectivity

You've traveled to lots of countries and consider yourself to be a competent Global Road Warrior. Best yet, you've got that new, state-of-the-art laptop computer that does everything.

But now you're in Berlin, Boston or Beijing and you can't quite get onto the Internet or retrieve your e-mail. Now what?

First of all, you're not alone. Even the most seasoned travelers have problems with mobile connectivity.

This chapter gives detailed solutions to the 20 most common problems people have with mobile connectivity.

Some problems are solved with a change in computer settings. Others require the purchase of a plug adapter or other piece of hardware. For hardware solutions see "Mobile Connectivity Suppliers" on page 114. You may also find some items at your local computer or electronics store. Many brand name products are available from general mobile connectivity suppliers such as iGo (www.igo.com).

20 Problems & Solutions Contents

Note:

In situations where a solution requires reconfiguration of software we have used Windows 2000 as the operating system example. If your computer uses a different operating system, the steps will likely be similar, but with slight variations.

Plugging in My Equipment/Appliances

1: My Electric Plug Does Not Match the Outlet

Different countries of the world may use different electric plugs and sockets than those you're used to at home. If your electric plug doesn't match the outlet in the country you are visiting, you cannot use your electric appliances and electronic device.

Solution: You Need an Electric Plug Adapter

The best way to get an adapter plug is to purchase one before you leave on your trip. You can find adapters online or at many travel and luggage stores.

If you forgot to purchase an adapter before departure, check with the airport's gift store upon arrival, as such stores often carry the required adapters. Also, check with your hotel, which might be able to provide you with one. Other than that, you can probably find an adapter in an electrical store in the country you are visiting.

✍ For a list of electric plug types see "World Electric Plugs" on page 105 and "Electric & Phone Plug Chart" on page 106.

✍ For a list of adapter suppliers see "Mobile Connectivity Suppliers" on page 114. See also "Electric Plug Adapters" on page 102.

Notes

_____

_____

_____

_____

_____

_____

_____

_____

_____

_____

_____

2: Appliance Does Not Use the Correct Voltage

Some countries use different voltage to power appliances and electronic devices. **Warning**: Connecting appliances and electronic devices to incorrect power outlets may damage or destroy your appliances or devices.

Solution: Use a Transformer or Converter

Transformers and converters take the electrical voltage from the electric outlet and alter it to be compatible with your electric appliances and electronic devices. The words transformer and converter seem easily interchangeable, but there are very important differences.

Converters Use a converter for heating appliances ONLY. These include: - hair dryers, curling irons, electric blanket, heating pads. Converters are used only for short periods of time (generally 10 minutes to 2 hours). Such converters are available for appliances with 50-1857 watt ratings.

☛ Be ABSOLUTELY certain that the wattage capability of the converter you purchase is sufficient to run your appliance. Look for the wattage rating on the label. Do <u>NOT</u> assume that "a higher wattage converter can handle a lower wattage laptop." Converters work on a totally different principle than transformers and can burn out electronics such as computers, cameras, and the like.

Transformers Use a transformer for BOTH *electronic devices* and *electric appliances*. Electronic devices include:
- computers, printers
- battery chargers
- any device with a computer chip
Travelers requiring continuous use of an appliance or device (continuous use for more than one or two hours) should bring along a "continuous use transformer," regardless of whether the appliance could be used with a converter.
A "2 to 1 step-down transformer" will take 220-volt electricity and enable you to run a 110 volt appliance.
A "1 to 2 step-up transformer" will take 110-volt electricity and enable you to run a 220-volt appliance. Transformers are HEAVY. We recommend that you either get a dual-voltage travel appliance or leave it at home.

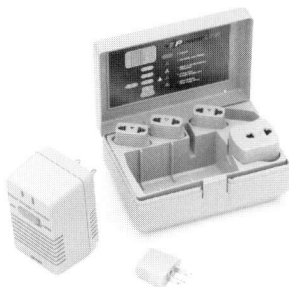

☛ A high-low combination with a low-voltage transformer and a converter for 50-1875 watts in a single unit is also available.

☛ Most laptop computers come with a transformer on the power cord that adapts 110-220 volt current to the required current of the computer.

Cautions

☛ When in doubt, use a transformer with the appropriate wattage.

☛ The wattage of your appliance or device must fall within the transformer's wattage range; to be on the safe side, give it a 10- to 20-watt buffer.

✍ For a list of suppliers, see "Mobile Connectivity Suppliers" on page 114.

3: What Size Converter or Transformer Should I Use?

It is important to use the correct device, or you could risk blowing out your appliance. Converters are available in a range of 50 to 1875 watts, and transformers in ranges of 50, 100, 200 watts or more.

Solution: Check the Wattage on Your Appliance

Look for the voltage switch or a label on your appliance to determine which size converter/transformer you need.

Low Wattage Appliances: (up to 50 watts) generally include radios, portable CD players, battery chargers, shavers, and contact lens cleaners. Do NOT use a high-wattage "converter" for electronics.

High Wattage Appliances: (50 to 1600 watts) generally include garment steamers, hair dryers, irons, heating pads, curling irons, and shaving cream heaters.

Transformers with a capacity greater than 200 watts are quite heavy. They are often more expensive than purchasing a second appliance with the needed voltage. A high-low combination with a low-voltage transformer and a converter for 50-1875 watts in a single unit is also available, but one must be very careful to place the switch in the "low" position before connecting any electronics to it.

✍ For a list of electric converter and transformer suppliers see "Mobile Connectivity Suppliers" on page 114.

Notes

_____

_____

_____

_____

_____

_____

_____

_____

_____

_____

_____

_____

_____

_____

_____

_____

_____

4: Planning for Power Surges and Outages

Power surges and outages can damage your computer hardware and other electronic equipment. Surges send a spike of electricity, which can burn out the hardware. Outages can cause you to lose data in open programs at the time of the outage.

Solution: Get a Surge Protector

To prevent power surges and outages from harming your equipment, get a surge protector. This will not keep your computer from shutting down during a power outage, but it will stop the spikes of electricity from causing severe damage. Check the voltage of the country to which you are traveling to ensure you get the right surge protector.

You may also need an electric plug adapter (Problem 1) and perhaps a converter or transformer (Problem 3) in order to use this equipment in a foreign country.

Computer, hardware, and electronics stores carry surge protectors but they are usually of the bulky office variety. Look for special pocket-sized surge protectors (shown at right) made specifically for the traveler.

☛ If you are still concerned about losing your computer work during an outage, save your document files often.

☛ Make sure that the intended voltage of the surge protector you use is the same as the one in-country. Do <u>NOT</u> use a U.S. (110v) surge protector on a 220-volt powerline unless this is done after the use of a 220-to-110v transformer, and, even then, avoid it because autotransformers used in most "50W" transformers sold will likely burn your U.S. surge protector. Similarly, do <u>not</u> use a 220v surge protector in 110v countries as, it will be ineffective.

✍ For a list of suppliers see "Mobile Connectivity Suppliers" on page 114.

Notes

_____

_____

_____

_____

_____

_____

_____

_____

_____

_____

Plugging in My Modem

5: Wall Outlet Does Not Match Modem Cord Plug

Different countries use different phone plugs and outlets. You may get to a hotel or office and find that you cannot plug your modem cable in because the plug does not match the outlet.

Solution: Get a Modem or Telephone Adapter

It is best to research ahead of time to establish which phone plugs are commonly used in the country to which you are traveling. You may need to purchase a phone plug adapter, often referred to as a modem or telephone adapter. A plethora of adapters exist, some specific to the country to which you are going, and others that serve multiple types of plugs. Go to a supplier Web site or product catalog to find one that suits your purposes.

✎ For a list of suppliers see "Mobile Connectivity Suppliers" on page 114.

6: How Can I Connect from a Pay Phone (or Other Hardwired Phone)?

In some hotels and offices you may find that there is no telephone wall jack. For example, 1) the phone may be hardwired into the wall, or 2) you may be at a public telephone and unable to plug in your modem.

Solution 1: Use a Digital Line Connector

If you can disconnect the handset from the body of the phone (and one often cannot), you can use a digital line connector. It plugs into the phone where the handset cord connects and works for both digital and analog phone systems. This is so because the signal between a telephone handset and the rest of the telephone is always analog, even if the telephone line to the digital telephone is digital.

Some digital connectors, such as the Modem Doubler by Road Warrior (shown at right), require two 9-volt batteries and may come with an optional AC power adapter. These adapters are not usually dual-voltage (Problem 2), so you may need a transformer (Problem 3) and possibly an electric plug adapter (Problem 1).

Solution 2: Use an Acoustic Coupler

When the handset is hard-wired to the phone (pay phones, for example), use an acoustic coupler. The coupler is a cradle, which holds the phone's handset and serves as a data transfer bridge to your computer. Place the phone's handset on the coupler and plug the coupler's cord into your modem port. The coupler now transmits the dial tone from the phone to your computer. The Coupler by Road Warrior is shown at right. Such devices are not recommended because:

1. They pick up a lot of sounds from the room, which results in data errors.

2. Since one cannot usually dial through the acoustic coupler, one has to become quite precise in timing the time between when a call is dialed by hand and when one's laptop is told (also manually) to handshake the modem connection.

✍ For a list of suppliers see "Mobile Connectivity Suppliers" on page 114.

Solution 3: Rewire Wall Connection

Some of the more experienced and brave global road warriors do some direct connects to analog phone lines at the wall connection if they are handy with tools and the wall plate removes easily.

Since this involves pulling wires out of the wall, and possibly stripping some of them, it is not recommended unless you really know what you're doing. You may end up tearing apart the phone system and causing damage, at which point the phones won't be working and you may be financially liable to the hotel. Often, there are more than two wires in the jack. If the "green" and "red" selections do not work (or cannot be identified), use a pair of scissors or any metallic object to temporarily short out any two wires and listen for a very audible "click" when you identify the two wires that you should connect to.

☛ Caution: Never attempt hard wiring to a Digital PBX telephone system. If you are unsure if the line is analog, test it using a telephone line tester (see Problem 7). Usually, a telephone with a number of feature buttons (e.g., message, etc.) is a digital phone, unless these feature buttons are merely pre-programmed telephone numbers (e.g., to call room service).

☛ Avoid disconnecting the room's telephone altogether, as some hotels are wired to detect this as an alert of theft of their telephone.

✍ For a list of suppliers see "Mobile Connectivity Suppliers" on page 114.

7: Are the Phone Lines Digital or Analog?

Computer modems use analog signals. Many hotels, offices, and universities, however, use digital telephone systems. These are often identified by a digital display or extra buttons for room service, etc., but it can be difficult to be sure.

Digital systems are unable to process analog signals (and vice versa without a modem), which makes it impossible for modems to transfer data. If you plug into a digital system, you may merely get a "no dial tone" message on your computer. Far worse consequences of plugging into a digital system include activating special operations of the system (for example, calling room service or triggering an emergency alarm), causing the digital system to malfunction, or completely ruining your modem. Because digital systems operate on a voltage level that is greater than most modems are designed to handle, PC card modems (commonly called PCMCIA modems) are extremely susceptible to damage.

Solution: Look for a Dataport or Get a Line Tester

First find out if the telephone has a dataport (analog jack/modem port). Four- and five-star hotels usually have them on their phones. You can call the hotels where you will be staying and the offices where you will be working to check. Also confirm the type of dataport, as it may not match the plug on your modem. You may need to purchase a telephone plug adapter (Problem 5). If a dataport does exist, that connection is almost always analog and you can connect your modem to it.

If no dataport exists, you will need to determine if the line is digital or analog. Purchasing a Line Tester, like the Modem Saver International by Road Warrior (shown above), will give you this answer if no one on the premises can. Merely insert the tester in the wall phone jack, and it will determine the status of the line.

☛ If in doubt and you must have an analog line, you can request a room with a Fax connection. All Fax lines are analog and safe to plug your computer modem into.

✍ For a list of suppliers see "Mobile Connectivity Suppliers" on page 114.

Notes

8: The Digital Phone Has No Dataport

Some hotels or offices operate on a digital system and do not offer analog dataports on their phones. (If you are unsure whether the phone is digital, see Problem 7.) In this case, you have two options:

Solution 1: Digital Interface/Line Connector

If the telephone is not hard-wired between the base and the handset, you can use a digital interface/line connector. This will give you an analog signal from the handset jack. Digital connectors like Konexx's Data Port Anywhere (shown at right) draw power directly from your laptop's USB port, meaning you don't have to carry batteries or an AC adapter. But, make sure that any line connector you purchase is compatible with your computer software.

Solution 2: Get an Acoustic Coupler

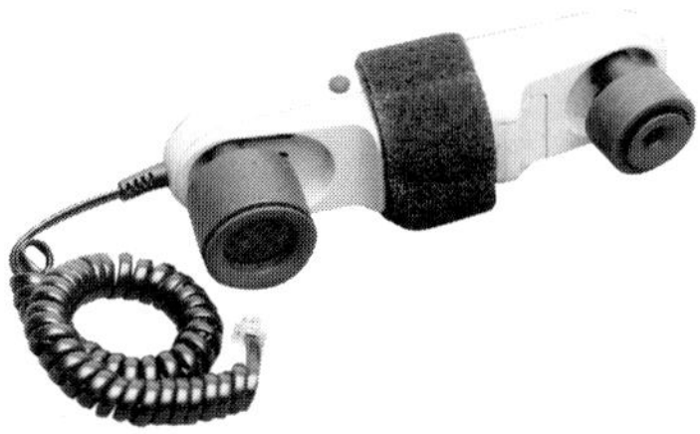

If the telephone is hard-wired between the base and the handset (meaning you cannot unplug the handset from the base), an acoustic coupler is required to physically attach the handset to the modem. Acoustic couplers, like the one from Road Warrior (shown at right), are generally battery operated. Acoustic couplers provide a safe connection through the handset of almost any kind of telephone: public pay phone, hard wired, or digital PBX.

✍ For a list of suppliers see "Mobile Connectivity Suppliers" on page 114.

Notes

9: Reversed Polarity or Telephone Wall Jack with Multiple Lines

9A) Reversed polarity is caused by incorrect wiring inside the wall jack. While a phone with reversed polarity may operate well when you make a voice call, your modem may perform at lower speeds.

9B) Modems are often set to work on a specific pair of wires in the wall jack (commonly the inner pair). When multiple lines come from the jack, your modem may end up operating on a set of wires it isn't designed for, and your modem won't work.

Solution for 9A: Line Tester

To check polarity, try the ModemSaver International by Road Warrior (shown at right), or another similar line testing product. If your line tester establishes that the polarity is reversed, you will need an adapter to correct it. The modem will then dial smoothly. The ModemSaver comes with an adapter for reversing polarity.

Solution 1 for 9B: Line Adapter

A line adapter will "switch" the lines so the modem believes it is operating on the correct pair of wires. ModemSaver International by Road Warrior comes with this device already installed.

✍ For a list of suppliers see "Mobile Connectivity Suppliers" on page 114.

Solution 2 for 9B: Connect Wires

Your final option is to pull the telephone receptacle from the wall and connect to the two wires of interest (identified by temporarily shorting every pair until you hear a distinctive "click" on the handset) with alligator clips

Notes

Configuring My Modem

10: Using Pulse/Rotary Telephones

Pulse, also known as rotary dialing, telephones have become obsolete in the United States and most of the world, but pulse telephones still exist. Pulse telephone *lines* can cause a problem when you attempt to dial in with your modem, and even more problems when attempting to call home to get your voice mail, reach someone's personal extension, or access your credit card and bank accounts.

Solution: Reconfigure Software

For your modem, it may be possible to set your software configuration to use pulse dialing when dialing out.

On Your PC (Windows 2000):

1. On the desktop, double-click on My Computer.
2. In that folder, locate the Control Panel Folder and double-click.
3. In that folder, locate the icon for Phone and Modem Options and double-click.
4. This will open a dialog box with three tabs at the top. Locate and click the Dialing Rules tab.
5. Select My Location and then click on the Edit button.
6. You should now have an Edit Location dialog box with three tabs at the top. Click the General tab and then locate the Tone/Pulse radio buttons at the bottom of the dialog box.
7. Select Pulse, then click OK to exit the dialog box.

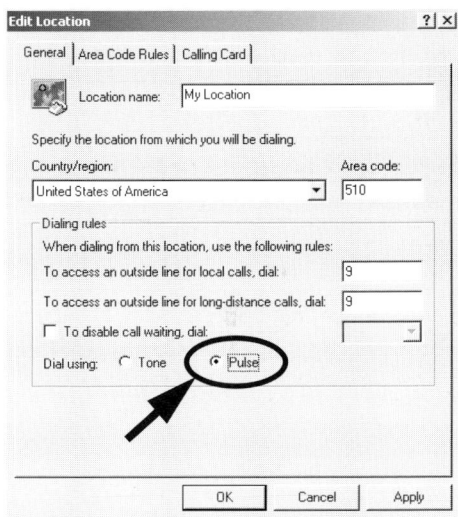

Notes

11: My Modem Does Not Recognize the Dial Tone

In other countries, the dial tone may not sound quite like the dial tone on your phone at home. The difference in tone may cause your computer to think there is no dial tone, thus, it won't want to dial a connection. Possible error messages from your computer might include: "No dial tone," "Line is busy," or "No carrier."

Solution: Edit Scripts or Configuration for Modem

To get your modem to disregard the odd dial tone, you can edit the modem scripts or turn off the modem's search for a dial tone. This may be a good idea, regardless, even in the U.S. where you can always listen-in either through your modem's built-in speaker or through a telephone connected to the same line.

To edit your modem scripts, contact your modem manufacturer, who can give you a script specific to your modem and to what you want to accomplish.

Following are examples of how to turn off the modem's search for the dial tone on a PC operating on Windows 2000 and on the later Macintosh operating systems.

On Your PC (Windows 2000):

1. On the desktop, double-click on My Computer.
2. In that folder, locate the Control Panel Folder and double-click.
3. In that folder, locate the icon for Phone and Modem Options and double-click.
4. This will open a dialog box with three tabs at the top. Locate and click the Modems tab.
5. Select your modem from the list by highlighting it and then click on the button labeled 'Properties' in the lower right corner.

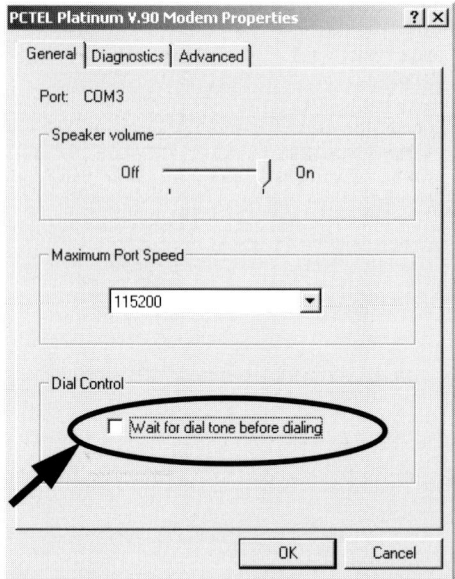

6. This opens a final dialog box. At the bottom you will notice a checkbox beside the notation: "Wait for dial tone before dialing." Uncheck this box by clicking on it.
7. Then click OK to exit the dialog box.

On the Macintosh:

If you are using TCP/IP and PPP or Remote Access, you cannot easily edit the modem scripts on your own. You may want to contact your modem manufacturer to see if they have a special driver (drivers allow communication between your operating system and your modem) for your situation. You could also read the Macintosh User Manual on Open Transport and Editing Modem Scripts, though this may be more than the novice Macintosh user is ready to do.

You can, however, turn off the modem's search for a proper dial tone:

1. Go to your Apple Menu in the upper left-hand corner of your screen.
2. Select the Control Panels folder
3. Within Control Panels, select the Modem control panel.
4. In this dialog box, you should be able to put a checkmark in the box that says "Ignore dial tone."

☞ If you are using "Config PPP" or "Free PPP" on your Macintosh, adding scripts should be similar to doing so in Windows (see above). Again, contact your modem manufacturer for exact scripts and where to place them.

Notes

12: My Modem Is Dialing Too Fast

Older telephone systems in some offices and hotels may not recognize numbers that are dialed by newer modems, as they dial too quickly.

Solution: Edit Init String

You can make your modem dial more slowly by editing/adding a command to the Init string. (Init string: an initialization string of code that your computer sends to your modem to tell it how to dial).

Windows 2000 Solution

1. Starting on the desktop, double-click on My Computer.
2. In that folder, locate the Control Panel Folder and double-click.
3. In that folder, locate the icon for your Network and Dial-up Connections and double-click.
4. This brings up a list of your connections.
   Choose the dial-up connection that you are currently using by clicking on it.

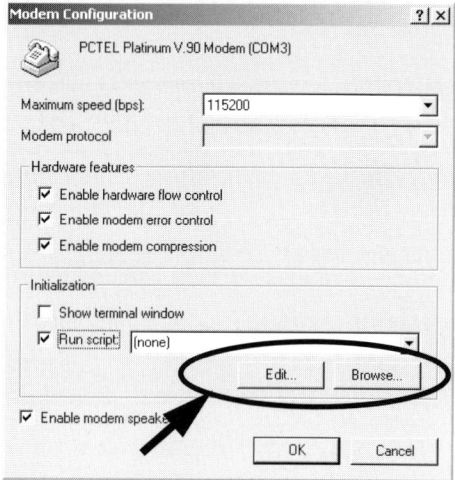

Now, right-click and choose Properties from the drop-down list.
5. A dialog box will open with five tabs, it defaults to the General tab. Find the Configure button on this tab and click.
6. Check the Run Script box by clicking it.
7. Under this tab, toward the bottom, find the button called Advanced and click on it.
8. This action will cause the Edit and Browse buttons to become active. You can choose to edit your Init file manually or browse for a script file (see below).

☛ To edit your modem scripts, contact your modem manufacturer first. They can give you a modem-specific or task-specific script which will work much better than a general script that we could provide.

13: I Need to Dial a Prefix to Get an Outside Line

When dialing out from an office or hotel, you may need to dial a number to get to an outside line ('9' is commonly used in offices in the U.S.). The same number is required for your modem to dial. However, merely typing the number in front of the number on your modem dial-up screen will not normally allow you to get an outside line since the modem will dial the prefix too rapidly.

Note: Before proceeding with the more elaborate procedure below, try entering the "outside prefix" number followed by one to three commas before the telephone number; this introduces enough delay to satisfy most existing telephone systems, e.g., 9,,001503. Make sure that if you are dialing an out-of-city or out-of-country number, that you have inserted the requisite additional prefixes.

Solution: Enter Number in Modem Configuration

1. Starting on the desktop, double-click on My Computer.
2. In that folder, locate the Control Panel Folder and double-click.
3. In that folder, locate the icon for your Phone and Modem Options and double-click.
4. You will see a dialog box with three tabs. The box defaults to the tab labeled Dialing Rules; in this tab, select My Location and click the Edit button below it.

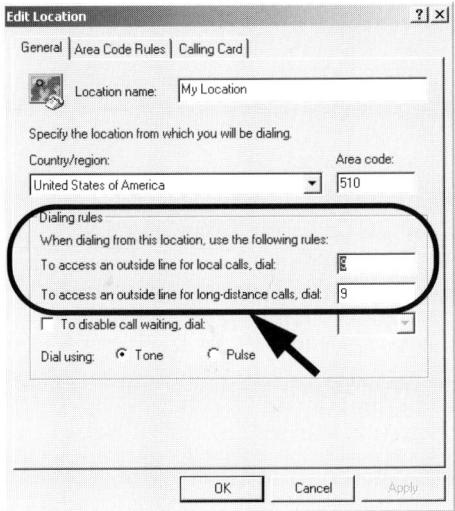

5. Another dialog box with three tabs will appear. Click the General tab.
6. In the Dialing Rules section of the General Tab enter the number required to access an outside line. You can now dial a local or long distance number.
7. Then click OK to exit the dialog box.

Notes

14: I Need to Use an Operator or Calling Card

When calling from hotels and some offices, you may need to make an operator-assisted or calling card call to dial out. Dialing from a modem is a one-step dialing process; consequently, the modem does not know how to handle an operator or a calling card (multiple dialed numbers).

Solution: Change Modem Configuration

To circumvent this problem, you must configure your modem with the proper information that will allow it to take into account the operator or calling card.

Operator-Assisted Call on Your PC (Windows 2000):

1. Starting on the desktop, double-click on My Computer.
2. In that folder, locate the Control Panel Folder and double-click.
3. In that folder, locate the icon for your Network and Dial-up Connections and double-click.

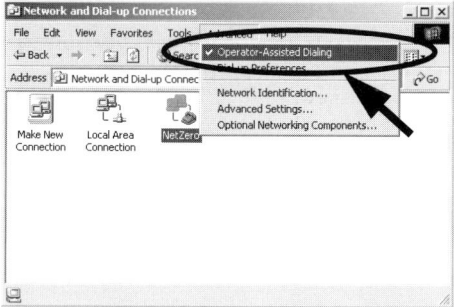

4. Find "Advanced" on the menu bar at the top of the window and click on it. Select Operator-Assisted Dialing.
5. A new dialog box will open, double-click the connection that you wish to dial.
6. Pick up the telephone handset, and dial the number or ask the operator to dial it for you (the number assigned to the entry is displayed in the dialog box for reference).
7. Immediately after you or the operator have finished dialing, click Dial; if you delay, the remote dialed host computer may hang up before your modem engages in the "handshaking" with it.
8. Hang up the handset only after the modem takes control of the line, which is typically signaled with a click followed by silence.

☛ It is always safe to replace the handset once Network and Dial-up Connections begins verifying your user name and password, or you risk room audio contaminating the data, resulting in errors and a failure to establish connection. The status message will remind you of this. If operated-assisted dialing is enabled, a checkmark appears next to Operator-Assisted Dialing on the Advanced menu.

Using a Calling Card on Your PC (Windows 2000):

1. Starting on the desktop, double-click on My Computer.
2. In that folder, locate the Control Panel Folder and double-click.
3. In that folder, locate the icon for your Phone and Modem Options and double-click.
4. You will see a dialog box with three tabs. It will default to the tab labeled Dialing Rules. Select My Location and click the Edit button below it.
5. Another dialog box with three tabs will appear. Choose the Calling Card tab and then select the type of card that you are using.
6. Now enter the relevant information to make your call.

☞ Remember to reset settings upon your return home!

☞ Hotels can add expensive surcharges when you dial directly from their telephone system, sticking you with an expensive bill at the end of your stay. You can avoid these excessive charges by using your calling card.

☞ Keep in mind that hotels maintain records of which number was dialed by each room and when.

Notes

15: My Modem Is Not Responding

Your computer may tell you that the modem is not responding. This can happen for different reasons. Read the error message carefully to determine the problem. During the first attempt to connect from a new location, it will help if you <u>quietly</u> listen-in on a phone connected to the same line. Did your modem dial out? Did the remote number you dialed ring? Did it answer with the characteristic modem tones? Did your modem respond? Often, this information is more useful in diagnosing a problem than anything that your computer's screen can display. In the vast majority of situations, you will usually find that you never really rang up the number you were calling because of some missing prefix in the dialing sequence; this is particularly true if your modem "worked" at home.

If this does not resolve the problem, then:

Solution 1: Reboot

Sometimes the modem simply does not initialize properly. Often, if you turn the modem off and then back on again, it will reset itself properly. Since most laptops have internal modems, you will have to shut down (NOT restart!) your computer, leave it off for about ten seconds, then boot back up again.

☞ Restarting is not the same as shutting down. It only reboots the software, not the hardware, which means it only reboots the operating system on the hard drive, not the actual components of your computer. You must shut down completely to reboot the modem and other hardware.

Solution 2: Check Dial-Up Networking (PC)

The following are general guidelines to help you get to screens on your computer that may need modification. You will most likely need to contact your Internet Service Provider (ISP) to confirm some of these steps and to get information specific to these screens. The sections below contain generic numbers that do not dial an ISP or give a proper configuration for your ISP.

1. Starting on the desktop, double-click on My Computer.
2. In that folder, locate the Control Panel Folder and double-click.
3. In that folder, locate the icon for your Network and Dial-up Connections and double-click.
4. This will bring up a list of your connections, choose the dial-up connection that you are currently using. Right-click on it and choose Properties from the drop-down list.

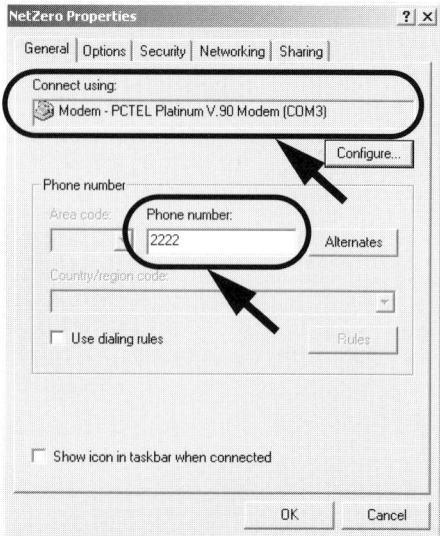

5. A dialog box will open. Check the items here to make sure they are correct.

6. At the top of the window is a tab called Networking. Click on it for more options.

7. Next locate the section labeled, "Type of dial-up Server I am calling." Click on the drop-down arrow. Ensure that the information you see matches your ISP's information.

8. Further down in the same dialog box find "Internet Protocol (TCP/IP)." Choose this component by clicking on the words (not the check box) and then click on Properties.

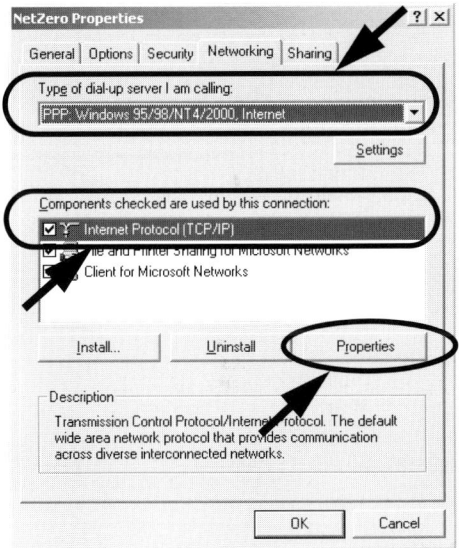

9. This opens another dialog box that allows you to enter the specific IP address and DNS server address that was issued to you by your ISP. More than likely, the values that would be entered in your TCP/IP settings are issued and placed automatically as soon as you connect to the Internet. All you have to do is configure your settings to obtain the IP address and DNS server address automatically by selecting the appropriate radio buttons.

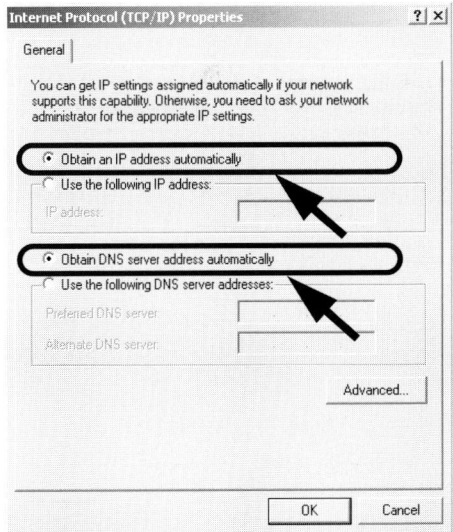

Solution 3: Check the Modem Control Panel (PC)

1. On the desktop, double-click on My Computer.
2. In that folder, locate the Control Panel Folder and double-click.
3. Then locate the icon for your Phone and Modem Options and double-click.
4. This opens a dialog box with three tabs at the top. Find the Modems tab and click on it.
5. A list of connections appears. Select your modem from the list by highlighting it and then click the button labeled "Properties," in the lower right corner.

PCTEL Platinum V.90 Modem Properties   ? X

General   Diagnostics   Advanced

Modem Information

Field	Value
Hardware ID	PCI\VEN_134D&DEV_7897&SUBSYS_00011

Command	Response

Query Modem

Logging
☐ Append to Log   View log

OK   Cancel

6. Another dialog box opens with three tabs at the top. Choose the Diagnostics tab. You can query the modem and read the modem log from this dialog box. These two items can help you discover what ails your modem.

Notes

_____

_____

_____

_____

_____

_____

_____

_____

_____

_____

_____

_____

_____

Problems Online

16: Tax Impulsing is Disconnecting My Modem

Tax impulsing is a signal of 12 KHz or 16 KHz metering pulses sent from a telephone company's office to a subscriber's phone during a call to detect the length of the call and establish how much to bill the subscriber.

Unfortunately, to your modem, tax impulsing is a form of line noise that can disrupt your data transmissions. Ask someone at your hotel or the office you will be visiting if their telephone system uses tax impulsing, and keep in mind that the answer you get may be wrong because most hotels don't know such details. One way to tell if a country you are in uses tax impulsing is to go to a kiosk and ask to place a call; if the proprietor looks at a counter at the end of your call to base your charges on, tax impulsing is likely to be used in that country.

Solution: Use a Line Filter

By plugging a line filter into the telephone wall jack and then plugging your modem cord into the filter, you can prevent the tax pulse from reaching your modem and disrupting the connection.

The ModemSaver International (shown at right) by Road Warrior (iGo Corporation) is a line tester, which also has line filtering capabilities. Another option is the TeleFilter (bottom right) by Teleadapt.

☛ Countries that use tax impulsing include Austria, Belgium, Czech Republic, Germany, Greece, Hungary, Italy, Slovak Republic, Slovenia, Spain, and Switzerland.

✍ For a list of suppliers see "Mobile Connectivity Suppliers" on page 114.

Notes

17: Line Noise is Disconnecting My Modem

Line noise is usually caused by a poor connection somewhere between your modem and the local telephone relay station. A possible source could be the line that connects your modem to the wall jack, but most likely it lies with the wiring inside the wall or the lines outside. Since, in older telephone systems, line noise is dependant on which particular circuit you happened to be connected to when you picked up the phone, try redialing and listen in to the connection for a few seconds from a handset on the same phone line.

If the problem persists, then consider the fix below:

Solution: Edit Settings on Your Computer

You can usually solve the problem by editing your modem settings to lower your connection speed and disable compression. This enables the modem to slow down and deal with the line noise.

On Your PC (Windows 2000)

1. Starting on the desktop, double-click on My Computer.
2. In that folder, locate the Control Panel Folder and double-click.
3. In that folder, locate the icon for your Network and Dial-up Connections and double-click.
4. A list of your dial-up connections comes up choose the one that you are currently using. Right-click on it and choose Properties from the drop-down list.

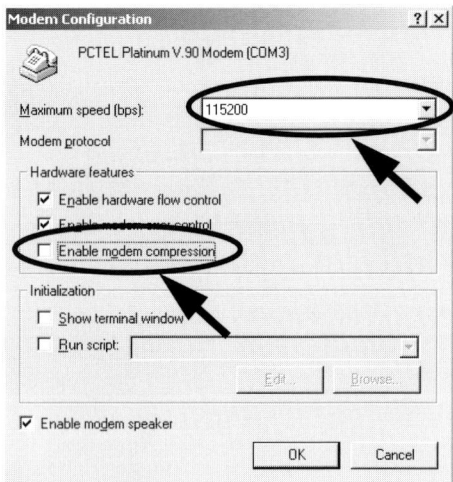

5. A dialog box will open with five tabs, which defaults to the General tab. On this tab find the Configure button and click on it.
6. Another dialog box labeled Modem Configuration will open. You will see your connection speed at the top of the box.
7. Choose a slower speed from the drop-down menu.
8. Next, uncheck the box labeled "Enable Modem Compression" located toward the bottom of the dialog box.

On Your Macintosh

You cannot change the speed on the new Macintosh control panels, but you can change the data compression.

1. Go to your Apple Menu in the upper left-hand corner of your screen.

2. Select the Control Panels folder.

3. Within Control Panels, select the PPP (Remote Access) control panel.

4. At the bottom, you will see an Options button. Click on it.

5. It will bring up a screen with 3 tabs across the top. The third tab should read Protocols. Click on it.

6. In this screen, you can uncheck the compression boxes.

Notes

_____

_____

_____

_____

_____

_____

_____

_____

_____

_____

_____

_____

_____

_____

_____

_____

_____

_____

18: Retrieving E-mail

When traveling, you may find that you cannot check your e-mail, or that you have to dial long distance to do so. Several easy and cost-effective solutions exist.

Solution 1: Web-Based E-mail

The simplest solution is to subscribe to free Web-based e-mail and use this as your traveling e-mail address. Many Web sites offer Web-based e-mail (www.yahoo.com, www.hotmail.com, and www.juno.com), and some even offer the option of sending and retrieving your e-mail over an encrypted (SSL) connection; examples include: www.cotse.net, www.safe-mail.net, and www.fastmail.fm.

You can access these services anywhere you can get a dial-up connection (an Internet connection). They do not require use of your own dial-up connection but allow you to check e-mail at Internet cafés or through an ISP that is local to your current location (anywhere that you can get Internet access). Be wary about entering your login name and password from untrusted locations, as they may be captured by these terminals. Opt, instead, for logging in after you have established a secure (SSL) connection; this does not guarantee protection from this threat, but goes a long way towards reducing the vulnerability.

Your Web-based e-mail address will not automatically receive e-mail that was sent to your work or home e-mail address. You will need to check with your ISP to see whether or not they will forward the e-mails addressed to those locations.

Solution 2: Mail Forwarding

Some ISPs will forward your e-mail, meaning that you can have your local e-mail from home or work forwarded to your traveling e-mail address (your Web-based e-mail address) while on the road. For example, if the e-mail address on your business cards reads your_name@company_name.com, you may be able to have any e-mail received at that address forwarded to your_name@yahoo.com (for example) while on your trip. Please consult with your ISP and make sure that you subscribe to the alternate traveling e-mail address before attempting mail-forwarding.

As another option, AOL has local dial-up numbers all around the world that allow you to access your AOL e-mail through their interfaces, as long as you have access to the Internet. However, you need to be sure that they do have local access in the country to which you plan to travel (refer to our country listings).

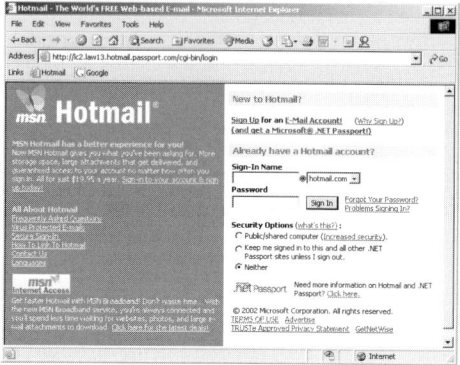

Miscellaneous Problems

19: Battery Life

Batteries naturally run out of power after an allotted time.

Solution 1: Carry Spare Batteries

Carry a spare battery and a battery charger so that a battery will always be charging. While spares and their chargers can be cumbersome, it is a good idea to have them, especially if your particular battery type is difficult to locate in stores. Even rechargeable batteries eventually wear out. Keep in mind that most rechargeable batteries lose their charge if subjected to a mechanical shock, so pack them well. Also keep in mind that NIMH (nickel metalhydride) batteries self-discharge quite rapidly within a week to a month; lithium ion batteries maintain their charge much longer.

Cell phones and some laptops let you recharge the battery while it is inside the equipment. You simply need a power cord that plugs into a wall jack or a car cigarette lighter. Having both is a good idea. You may need an adapter, as foreign electric outlets may be different (Problem 1).

Solution 2: Get a PowerXtender

Another option for your laptop is the PowerXtender (shown at right), manufactured by Xtend. PowerXtender allows you to plug into cigarette lighters or the power outlets found on many airlines in business- and first-class cabins. Make sure the PowerXtend works with your laptop before purchasing

☛ As yet, very few airline coach seats offer the option to recharge one's computer in flight. Recharging it from the 110v outlet in the lavatory is an impractical option.

✍ For a list of suppliers see "Mobile Connectivity Suppliers" on page 114.

Notes

20: Laptop Computer Security

Laptop computers are one of the most frequently stolen items from business travelers. Be careful in airports and other crowded areas, especially if you use an easily recognizable "laptop bag." However, leaving it in a hotel or office may prove no safer.

Solution: Purchase an Antitheft Device

Some hotels have room safes, but those may only be large enough to hold your smaller valuables. As such, you may have to rely on other means to guard your laptop, whether it is in your hotel room or lying next to you on a bench at the airport.

The Kryptonite motion sensor alarm, by iGo, allows you to secure valuables such as your laptop for about US$60. A laptop security adapter connects the alarm cable directly to your computer, and a flashing LED display will warn thieves that the alarm is activated. And if that still doesn't stop them, a surround-sound alarm with a 110-decibel blast will startle the thief, and alert you to the infraction.

Various key cable security locks and other antitheft devices, such as the Kryptonite Key Cable lock (shown right), are also available from iGo and other suppliers.

It is preferable to store your computer into a locked suitcase while away from your hotel room. If data theft is the primary concern, become adept at removing your laptop's hard disk and taking it with you when leaving the room. This is so because a prominently displayed laptop "secured" with a cable such as depicted above is both inviting and easy to steal by cutting the cable.

☛ Always be aware of your surroundings and keep valuables out of sight when possible.

✍ For a list of suppliers see "Mobile Connectivity Suppliers" on page 114.

Notes

Glossary

Computer/Wireless/Internet

Notes

1. This Glossary contains the most commonly used terms in the world of computers, the Internet and wireless technology. We have sought to include comprehensive definitions and explanations, especially of terms that are poorly understood and, as a result, often misused. Overly technical terms have been omitted.

2. Many terms are listed by their acronym when that is the more common usage. For example: ISP, rather than Internet Service Provider. If the full term you are looking for is not listed, look for its acronym.

3. The terms "wireless," "cellular," and "mobile" are used interchangeably to refer to communication equipment that transmits and receives by means of radio transmissions.

1G Wireless
(wireless) First Generation Wireless Technology. Technology that permitted the user to make simple phone calls.

2G Wireless
(wireless) Second Generation Wireless Technology. Wireless technology that permits the user to make phone calls, record and play back voice mail and receive simple e-mail messages. This is the most common wireless technology available as of early 2002. Approximate data transmission speed of 10 Kbps (kilo bits per second).

2.5G Wireless
(wireless) Second and a Half Generation Wireless Technology. Wireless technology that permits the user to make phone calls, send faxes, record and play back voice mail, send and receive large voice mail messages, do basic Web browsing, download simple maps and GPS navigation. This is the best technology available as of early 2002. Approximate data transmission speed of 64-144 Kbps (kilo bits per second).

3G Wireless
(wireless) Third Generation Wireless Technology. Wireless technology that combines a mobile phone, palmtop or laptop computer and television. This technology permits the user to make phone calls, send faxes, record and play back voice mail, send and receive large voice mail messages, do high-speed Web browsing, download maps, GPS navigation, conduct two-way video conferencing, TV streaming and electronic meeting planning and personal calendar. 3G Wireless promises to feature high worldwide commonality of design and compatibility of services. Approximate data transmission speed of 144 Kbps (kilo bits per second) to 2Mbps (mega bits per second).

10 Base-T
An Ethernet LAN (local area network) data connection over a twisted-pair cable with a transmission speed of 10 Mbps.
10—Refers to a media transmission speed of 10 Mbps (mega bits per second)
Base—Refers to baseband signaling which means that Ethernet signals are the only signals carried over the media.
T—Refers to a twisted-pair cable type.

100 Base-T
An Ethernet LAN (local area network) data connection over a twisted-pair cable with a transmission speed of 100 Mbps.
100—Refers to a media transmission speed of 100 Mbps (mega bits per second)
Base—Refers to baseband signaling which means that Ethernet signals are the only signals carried over the media.
T—Refers to a twisted-pair cable type.

1000 Base-T

An Ethernet LAN (local area network) data connection over a twisted-pair cable with a transmission speed of 1000 Mbps.

1000—Refers to a media transmission speed of 1000 Mbps (mega bits per second)
Base—Refers to baseband signaling which means that Ethernet signals are the only signals carried over the media.
T—Refers to a twisted-pair cable type.

access fee

(wireless) A monthly charge for connection to a wireless (cellular) network. Most access fees are now bundled as part of a service plan that includes both access and air time for a set monthly fee.

access provider

See ISP (Internet Service Provider).

activation

(wireless) The set-up and configuration of a wireless phone enabling it to send and receive calls.

activation fee

(wireless) A one-time charge for activation of a wireless phone.

air time

(wireless) Chargeable time for making or receiving wireless phone calls.

AMPS

(wireless) Advanced Mobile Phone Service. The standard analog wireless phone service system operating at 800 MHz. AMPS technology permits 40 to 50 conversations per cell, poor privacy protection, and poor noise immunity. See DAMPS.

analog

The representation of data in continuous quantities, as opposed to digital, which is the representation of data in discrete units. The most common explanation of analog vs. digital uses clocks. An analog clock has hour, minute, and second hands that display time in a continuous manner. The information displayed is in a constant state of change. A digital clock, on the other hand, displays time in discrete units such as hour, minute, and second. It shows one second, then the next, and then the next.

(wireless) A voice transmission characterized by continuous change and flow achieved by modulating its frequency and amplitude. Analog wireless service is being replaced by digital, because digital signals are more secure, transmit and receive signals with less noise and interference, and importantly, are faster and more reliable for data communications, which is the future of wireless.

analog phone

(wireless) A wireless phone that carries analog (as opposed to digital) signals.

analog signal

(wireless) A radio signal characterized by continuous change and flow achieved by modulating the frequency and amplitude of the signal.

AOL

(Internet) Acronym for America Online. An online information service that combines access to the Internet, e-mail, databases, and other features to more than 20 million subscribers worldwide. Connect at www.aol.com.

application

(computers) Short for "application program." Computer software designed to perform a specific function directly for the user. An application typically manipulates data and then supplies results to the user. Examples of common applications include: word processors, spreadsheets, accounting and database programs. Web browsers and graphics software are also considered to be applications. Compare to "systems software," which includes programming languages and operating systems that are required to run applications software.

attachment (e-mail)

(Internet) A text, graphic, audio, video, or other data file that is sent along with an e-mail on the Internet. In most e-mail systems one selects "Attach" from a menu and then selects a specific file or document to attach. Attachments are often compressed when the file size is large.

AVI
Acronym for Audio Video Interleaved. A multimedia (audio/video) technology developed and defined by Microsoft for use on personal computers using the Windows operating system. In AVI technology, standard audio and video frames are stored in alternate (interleaved) pieces to provide animation at 15 fps (frames per second) at 160 x 120 x 8 resolution. Audio is provided at 11,025 Hz, 8-bit samples. AVI files require AVI player software, which may come with a browser or be downloaded from Microsoft. Other multimedia technologies in common use include MPEG and QuickTime. The AVI file extension is *.avi.

B2B
(commerce/e-commerce) Acronym for business-to-business. Commerce in goods, services, or information that takes place between business enterprises. Contrast to the exchange of goods, services, or information between businesses and private individuals (B2C or business-to-consumer). Business-to-business is a term associated with the pre-Internet "old" economy whereas B2B is a term associated with the "new" Internet economy. B2B sales on the Internet are expected to grow at a faster rate than the B2C sector.

B2C
(commerce/e-commerce) Acronym for business-to-consumer. Commerce in goods, services, or information that takes place between a business enterprise and a private individual. Contrast to the exchange of goods between companies (B2B or business-to-business). Business-to-consumer is a term associated with the pre-Internet "old" economy, whereas B2C is a term associated with the "new" Internet economy.

backup
(computers) A duplicate copy of a computer file such as a company database or application program. A backup is made for security purposes in case of loss of data because of an equipment failure, fire, theft, or other causes. Many businesses keep backup copies of important files off-site in a secure location such as a bank safety deposit box. Also used as a verb: to backup a file.

band
(wireless) A range of radio frequencies between two defined limits that are used for a specific purpose.

bandwidth
(computers/Internet) The data transmission capacity of an electric communications channel or connection. The more bandwidth, the more data can be transferred in a given period of time. Bandwidth is measured in bits per second (bps).
The technical definition of bandwidth refers to the difference between the highest and lowest frequency that a connection can transmit and is measured in hertz (Hz) or cycles per second. The greater the difference between these two values the more data can be sent in a given period of time. Bandwidth differs greatly depending upon a number of factors. LANs (local area networks) used in many office environments are much faster than WANs (wide area networks) such as dial-up modems, DSL, and T1-3 lines as used in most Internet connections. The following chart gives a sampling of bandwidth for the most common network configurations.

WAN Connections—Bandwidth
Switched Services
Dial-up modems: 9.6, 14.4, 28.8, 33.6
    and 56 Kbps.
ISDN: BRI 64-128 Kbps
      PRI 1.544 Mbps
Unswitched Private Lines
T1: 1.544 Mbps
T3: 44.7 Mbps
DSL: 144 Kbps to 52 Mbps.

LAN Connections—Bandwidth

Ethernet (10 BaseT): 10 Mbps
Fast Ethernet (100 BaseT): 100 Mbps
Gigabit Ethernet: 1,000 Mbps
Token Ring: 4, 16 Mbps
ATM: 25, 45, 155, 622, 2,488 Mbps+

banner ad

(Internet) A graphic image of generally 460 pixels wide by 60 pixels high (also 460 x 55 or 392 x 72 pixels) used on a Web page to advertise a product or service. Many banner ads are now animated and/or hot linked to an advertiser's Web site.

base station

(wireless) The physical location and equipment required to connect a cellular phone in a wireless cell to the network. The equipment includes a low-power radio transmitter/receiver, tower, antenna, and power supply. Also called a cell station.

baud

(computers/Internet) a) (common usage) A measure of data transfer speed of a computer modem expressed as the number of bits the modem can send or receive per second. b) (technical) A unit of signaling speed related to the number of times per second that the carrier signal can shift value, expressed as the number of state-transitions or symbols that can be transmitted per second on a connection. This value can be different from the number of bits that can be moved per second. For example, a modem running at 300 baud can move 4 bits per baud or 1,200 bps. The true technical baud rate depends on the type of connection. See bandwidth.

binary

(computers) a) Two parts or things. b) A base-two system of representing data. In modern computing, all data (text, images, audio, and video) is stored, read, and processed by the computer in the form of ON and OFF electrical impulses. This binary system (ON or OFF) can also be represented as 1 (ON) or 0 (OFF), or as TRUE (1) or FALSE (0). Each byte of data is represented by a series of these ON, OFF electrical impulses. For example, the letter "A" is "01000001." The representation of data in binary form is the basis of modern digital computers, digital audio recordings, and digital wireless communications. See bit, byte.

bit

(computers) Contraction of binary digit. The smallest element of data storage on a computer. In a binary system (0 or 1, OFF or ON) a bit represents either 0 or 1, OFF or ON.

Bluetooth

(wireless) A short-range wireless technology designed to network mobile phones, pagers, computers, and PDAs (Personal Digital Assistant) that contain a special microchip. Bluetooth enables individuals to coordinate their mobile and fixed position computing and communications devices within a 10-meter range. For example, Bluetooth technology enables a user to connect a mobile phone with a home or laptop computer and access the Internet via the phone. Bluetooth technology was originally developed by Ericsson, IBM, Intel, Nokia, and Toshiba.

bookmark

(Internet) A hyperlink marker that serves as a shortcut to a particular Web site or a specific page of content within a Web site. Web browsers enable users to automatically bookmark favorite Web sites or pages for quick retrieval at a later date.

boot

(computers) To start a computer and its operating system software. To start application software, the term "launch" is used; as in "launch Microsoft Word."

bps

(computers) Acronym for bits per second. See bandwidth, bit.

bricks and clicks
(Internet) A business that combines a physical location open to the public (bricks and mortar) with an Internet e-commerce presence (click of a mouse).

bricks and mortar
(Internet) A business with a physical location (building) open to the public, especially retail locations, as opposed to a business that operates only on the Internet.

broadband
(computers/Internet) High-speed data transmission over a network. Broadband is generally considered to be a speed of 1.544 Mbps (T1 line) or faster. *See* bandwidth, DSL, ISDN, T1-3.

browser
See Web browser.

byte
(computers) Contraction of binary table. A unit of computer storage that consists of 8 bits and that generally represents a single character (a, b, c, 1, 2, 3, $, %, etc.). Computer storage (either in RAM, disk space, or file sizes) is generally stated in either kilobytes (KB or thousands of bytes), megabytes (MB or millions of bytes), or gigabytes (GB or billions of bytes). See bit.

call blocking/screening
(wireless) An optional phone service that automatically rejects calls from a user-defined list of telephone numbers.

call forwarding
(wireless) An optional phone service feature that automatically transfers incoming calls to another phone number of the user's choice. Call forwarding can be activated and deactivated simply by entering a code into the handset.

call waiting
(wireless) An optional phone service feature that notifies a user of an incoming call while another call is in progress, and allows the user to answer the second call while the first call remains on hold.

caller ID
(wireless) An optional phone service call-screening feature that displays a caller's phone number and, sometimes, the caller's name on the phone's display screen before the call is answered.

carrier
(wireless) A company that provides telephone, wireless, and/or other communications services.

carrier code
(telecommunications) A prefix used in certain countries to route a call through a specific carrier.

CDMA
(wireless) Code Division Multiple Access. A digital wireless communications technology that allows many users to share the same radio frequencies. For a transmission, the outgoing conversation is digitized and then tagged with a code. When the wireless phone receives a call it uses the code to pull only the specified conversation out of the many conversations on that frequency. CDMA technology is like a German-speaking person who is in a large crowded room of Dutch-speaking people but who can hear another German-speaking person across the room.

CDPD
(wireless) Cellular Digital Packet Data. An enhanced analog cellular technology that breaks data transmissions into "data packets" that are small enough to be sent on idle cellular voice networks. CDPD technology is like breaking a freight shipment into several smaller shipments in order to fit and fill leftover space in a number of cargo containers.

CD-ROM
(computers) Compact Disk-Read Only Memory. A storage medium using the compact disk format and used to record data, text, graphics, and/or audio files. The most common data capacities for CD-ROM disks are 650 and 700 MB. Note that computer CD-ROM drives can play both CD-ROM and audio CDs but that audio

CD players can play only audio CDs. CD-ROMs provide more stable data storage than floppy disks and other magnetic storage mediums and are used extensively for the distribution of computer software, games, and other data.

cell
(wireless) The geographic and spatial area served by a single wireless cell site. Each cell is equipped with a low power radio transmitter/receiver. Cell size is a function of the wireless technology used, terrain and capacity demands. As a wireless phone user travels from one cell to another the call is handed off to the new cell.

cell station
(wireless) The physical location and equipment required to connect a cellular phone in a wireless cell to the network. The equipment includes a low-power radio transmitter/receiver, tower, antenna, and power supply. Also called a base station.

cellular
(wireless) A wireless communications system and technology that divides a service area into a multitude of cells, each with a cell site (base station). Cellular calls are transferred from cell site to cell site as the user travels from cell to cell.

cellular phone
(wireless) A wireless radiotelephone that operates within the service area of a cellular network.

centrex
(telecommunications) A subscriber-owned and located telephone network (PBX or private branch exchange) that is connected to outside lines and telephone network, but where switching occurs at the local telephone office rather than at the organization's offices.

CGI
(Internet) Common Gateway Interface. A methodology by which a Web server acts as the interface for a Web user's request for information from a database. Essentially, the Web server receives a Web user's request, interfaces with an application program to query a database and then forwards the result back to the user. For example, a Web user might enter data on an online application form. This data is sent to the Web server, but is then forwarded to an application program which processes the data and sends a confirmation back to the Web server and finally back to the Web user. This is called server-side scripting and is a dynamic client-server application. The most popular scripting language for CGI is PERL. CGI scripts remain in wide use, although newer server-side technologies such as JSP (Java Server Pages), ASP (Microsoft's Active Server Pages), and freeware such as PHP are gradually replacing it, especially on Windows and Linux operating systems.

channel
(wireless) An individual segment of a radio frequency band along which a voice or data transmission is sent and received.

chip
(computers) A set of miniaturized electronic circuits used to process or store data. There are many types of chips designed for specific functions. Chip, microchip, and integrated circuit are used interchangeably.

cHTML
Acronym for compact HTML. A subset of HTML that is intended for formatting content displayed on small information devices, such as smart phones and PDAs, that have limited memory, low power, small storage, small screens and restricted input methods. cHTML does not support certain types of graphics as well as some advanced HTML elements such as tables, background colors, frames and style sheets. CHTML is in competition with WML (Wireless Markup Language).

client-server
The relationship between two computer programs within a single computer, or, more importantly, between two computers in a network (such as the Internet), where one computer (the client) makes a service request of another computer (the server) which fulfills the request. For example, a person may wish to check

the balance of a savings account from home using his or her computer over the Internet. In this case a "client" program in the individual's computer makes a request of a "server" program at the bank which retrieves the information from the bank's database.

client-side
Computer code that is executed on an individual user's computer and that does not involve sending a request to a server. Javascript, for example, is a client-side scripting language used to add dynamic capabilities to web pages.

compression
(computers) The encoding of data to reduce storage space. There are different compression methods and programs but they all work on the same basis of finding repeatable patterns of binary data (0s and 1s) and replacing them with a code and marker system. Compression is especially popular in sending large files (attachments) by e-mail. WinZip (www.winzip.com) is one example of a popular compression utility.

convergence
(wireless communication) The merging of various wireless, computer, satellite, internet and software technologies to produce a single device with worldwide capabilities as a mobile phone and PDA with Internet access.

(information technology) The combining of personal computers, telecommunications and television into a single user experience. For example, using a personal computer and the Internet to get video-on-demand.

CompuServe
(Internet) An online information service that combines access to the Internet, e-mail, databases, and other features. CompuServe is one of the oldest online services and is geared to business users, especially in the airline, insurance, and legal industries. CompuServe was acquired by AOL (www.aol.com) in 1998. Connect at www.compuserve.com.

cookie
Information stored on a user's computer (client) as a small text file by a Web site (server) that indicates what Web pages the user has visited and other user/client data such as preferences and passwords. Cookies enable a Web site to monitor and exploit the patterns and preferences of visitors to the site.

One user advantage of cookies is the automatic entry of passwords and other user data on HTML forms. However, cookies can accumulate and act as a history of the user's Web surfing, which can be used by commercial sites for user profiling. Also, anyone using a computer on which other users' cookies have been stored can access protected sites through cookie-driven passwords (this can be a problem particularly with shared laptops). Browsers such as Internet Explorer and Netscape allow the user to modify privacy settings and control the creation of cookies on the computer. Users can also search for a folder named "Cookies" on the computer's hard drive, the contents of which can be pared or deleted.

coverage area
(wireless) The geographic area served by a wireless telecommunications carrier. The same as service area.

crash
(computers) An event that renders a computer temporarily inoperable. Most computer crashes are the result of 1) an overload of instructions or data, 2) a conflict in software, or 3) a hardware failure. Most computer crashes can be resolved by rebooting (restarting) the computer. Also called an abend (abnormal end).

cross platform
(computers) A computer application (software) that is able to operate on all or a variety of different computer hardware or software platforms. The ability of an application to run across various platforms has become a hot topic in the age of Internet and wireless technology.

cross talk
(wireless) Telecommunications signal leakage from one wireless channel to another resulting in noise and distortion.

cyberspace

(Internet) A popular term used to describe the aggregate of information available on the Internet and other computer networks. The term was originally coined by William Gibson in his 1984 novel *Neuromancer* and (interestingly) referred to a futuristic computer network comprised of the minds (not computers!) of all the people who were "plugged" in.

DAMPS

(wireless) Digital Advanced Mobile Phone System. A digital wireless phone service system operating at 800 MHz. Technically, a proprietary TDMA (Time Division Multiple Access) technology. DAMPS technology allows for about 125 to 300 conversations per cell, good privacy protection, and high noise immunity. See AMPS.

default

(computers) The preestablished setting of a piece of computer hardware or software. For example, the default opening screen for a word-processing application might be a "New" document. Many default settings can be changed through user customization, often from a "Preferences," "User Preferences," or "Settings" menu.

digital

(computers) a) The representation of data in numerical form, usually in a binary system. b) Anything relating to numerical systems.

digital cellular

(wireless) Wireless technology that sends and receives signals in digital, usually binary, form. Digital cellular is the latest generation of wireless technology and provides the user with a number of advantages over analog systems, including better sound quality, reduced noise, fewer dropped calls, longer battery life, and greater privacy. Also, digital systems are able to handle many more calls than analog technology.

digital phone

(wireless) A wireless phone that sends and receives digital rather than analog signals. See digital cellular.

digital subscriber line

See DSL.

discussion group

See newsgroup.

domain name

See Internet domain name.

dot-com

(Internet) a) The period "." followed by the commercial (com) domain of an Internet domain name or e-mail address. b) Anything related to the Internet industry or economy. c) A company that does business on the Internet.

down

(computers/Internet) a) A computer that has not been turned on or that is not in operational condition. b) A Web site that is not operational. *See* up.

download

(Internet/computers) The transfer of computer documents or data files from a remote (host) computer to a local (client) computer over a network. To download means to receive and to upload means to transmit. *See* upload.

DSL

(Internet) Digital Subscriber Line. A technology that dramatically increases the data transmission capacity of telephone lines into homes and offices. DSL is one of the most popular forms of Internet access because it provides "always on" operation and uses existing telephone lines. DSL speeds, however, are a function of the distance between the subscriber and the telecom central office. The greater the distance, the slower the speed. There are many versions of DSL, but the two main groupings of service are Asymmetric DSL (including ADSL, RADSL, G.Lite, and VDSL) and Symmetric DSL (including HDSL, HDSL-2, SDSL, and IDSL). Asymmetric DSL features fast download and slow upload of data and is used primarily for Internet connections. Symmetric DSL features fast speeds in both directions but is used only where the distance from the telecom office is short. See bandwidth, E1, E3, ISDN, modem, T1-3.

dual band
(telecommunications) A mobile phone (handset) that is capable of sending and receiving signals in both 800 MHz cellular and 1900 MHz PCS frequencies. *See* dual mode, PCS.

dual mode
(wireless) A mobile phone (handset) that is capable of operating on analog and digital wireless networks. *See* dual band.

E1, E2, E3
(Internet, European standard) A dedicated digital circuit connection leased from a telecommunications provider that provides 2.048 Mbps (E1), 8.448 Mbps (E2), or 44.736 Mbps (E3) data transmission capacity. These lines are used for high speed private networks and connections to the Internet. E3 lines can handle full-screen, full-motion video. See T1-T3.

e-business
(Internet) Short for electronic business. An umbrella term for doing business online. In addition to simply processing transactions online (e-commerce), e-business includes all aspects of conducting business online including: buying, selling, marketing, advertising, order tracking, shipment tracking, and customer service. *See* e-commerce.

e-commerce
(Internet) Short for electronic commerce. Business transactions conducted on the Internet. The sale of goods or services online. E-commerce is a subset of e-business and generally refers specifically to the ability to process a transaction online. E-commerce also includes electronic data interchange (EDI) which is the structured exchange of business documents (e.g., inquiries, purchase orders, invoices, compliance documents, etc.) between computers.

EDGE
(wireless) Enhanced Data GSM Environment. A faster variation of GSM (Global System for Mobile Communications) wireless communications technology designed to transfer data at rates of 384 Kbps. This faster speed enables the delivery of multimedia e-mail and video conferencing applications to wireless terminal (cell phone) users. EDGE builds upon existing GSM standards, using TDMA (time-division multiple access) frame structure and existing cells. EDGE is a component of the migration of GSM and TDMA to 3G UMTS.

e-mail
(Internet) Short for electronic-mail. The transmission of electronic messages over a network from one computer to another, especially over the Internet. *See* attachment, spam.

EMS
(wireless) Enhanced Messaging Services. An open standard 3G wireless technology that allows users to send and receive text, melodies, pictures, sounds and animations as messages between EMS compliant GSM phones. EMS works on all GSM networks. EMS is an adaptation of SMS (Short Message Service). See also SMS.

encryption
(computers) The encoding of data for security while in transmission or at rest.

ESN
(wireless) Electronic Serial Number. A unique alpha-numeric code that is built into each cellular phone. Cellular providers use the ESN to activate a phone and properly track usage.

Ethernet
(computers) The most popular LAN (local area network) technology currently in use. Many PCs, most peripherals (such as laser printers), and all Macintosh computers come standard with an Ethernet port (RJ-45 connector) for connecting to a LAN or to a DSL or cable modem for Internet access.
Ethernet is available in three bandwidths:
Ethernet: 10BaseT (10 Mbps)

Fast Ethernet: 100BaseT (100 Mbps)
Gigabit Ethernet: 1,000BaseT (1,000 Mbps)
See bandwidth.

extranet
(Internet) A Web site with limited access, designed and maintained specifically for communicating with clients and customers rather than with the general public. An extranet uses the Internet as its delivery system, but restricts access via passwords. An extranet can be used to deliver vendor inventory information, specialized content such as research, or any other private data. Access to an extranet site may be free or on a paid subscription basis.

FAQ
(Internet) Acronym for Frequently Asked Questions. A list of commonly asked questions, with answers, maintained on a Web site to assist users with fundamental content, navigation, and usage information. It is considered a breach of netiquette (etiquette of the Internet) to call or e-mail asking a question that is answered in the FAQ section.

FDMA
(wireless) Frequency Division Multiple Access. Original AMPS analog cellular technology from the 1980s that has largely been replaced by newer digital systems. With FDMA, each of 30 available channels in a cell can only be assigned to one user at a time. Newer technologies, such as D-AMPS (Digital-Advanced Mobile Phone Service) use FDMA, but adds TDMA (Time Division Multiple Access) to enable three simultaneously calls per channel.

fiber optics
(telecommunications) Wire or cable made from glass fiber and designed to transmit digital signals (audio, video, or other data) as pulses of light.

firewall
(Internet) Computer software, or more commonly, a combination of computer hardware, software, and security measures designed to keep a computer or computer network secure from intruders. *See* hacker, virus.

flame
(Internet) A critical, often abusive message that is delivered by e-mail. For example, a flame may be received by someone who has abused the netiquette (etiquette of the Internet) by using public forums for advertising.

frequency
(wireless) A measure of an electromagnetic field (a wireless signal) in free space (the air) expressed as the number of complete cycles per second of the oscillating or varying current. The standard unit of measure is the hertz (Hz). If a current completes one cycle per second the frequency is 1Hz; if the current completes 60 cycles per second the frequency is 60Hz.

frequently asked questions
See FAQ.

FTP
(Internet) File Transfer Protocol. A standard protocol for transferring computer files over the Internet. FTP is often used by Web developers to upload their files to a host server. An FTP transfer is generally performed with a utility program or from within a Web browser. *See* upload.

GB
See gigabyte.

GIF
Graphic Interchange Format. A bitmapped graphics file format that supports 8 bit color (256 colors). The GIF format is the most widely used for displaying graphics on Web pages because of its ability to compress file sizes. In the process, however, certain characteristics of high-resolution graphics, especially photos, are lost or distorted. When Web masters wish to present images in high resolution, they often convert images to the JPG file format. Macintosh users refer to GIF files as "giff" files, while PC users refer to them as "jiff" files.

gigabyte (GB)
(computers) One billion bytes. *See* byte.

Global Positioning System
See GPS.

Global System for Mobile Communications
See GSM.

GPRS
(wireless) General Packet Radio Services. An enhancement to GSM cellular phone technology that enables users to make calls and transmit data (such as e-mail, Web browsing and file transfers) at the same time. This is achieved by the use of "data packets" which increased data transmission rates to as much as 114 Kbps. When introduced in the late 1990s, GPRS increased data transmission speeds by a factor of three in comparison with landline telephone networks, and by a factor of 10 in comparison with the existing Circuit Switched Data services on GSM cellular networks.

GPS
Acronym for Global Positioning System. A worldwide satellite-based radio navigation system consisting of 24 Earth orbit satellites operated by the U.S. Department of Defense and used to provide three- and four-dimensional position, time, and velocity information to military and civilian users who have GPS receivers. The system has two levels of service. **Standard Positioning Service** is available to anyone in possession of a GPS receiver (available for as little as US$100) and allows the user to establish their Earth location in latitude and longitude within a proximity of 10 meters. **Precise Positioning Service** is a highly accurate military positioning, velocity and timing service available only to the U.S. military and other authorized users.

How GPS works: Each of the 24 satellites contains a computer, atomic clock, and radio. The strategic positioning of these satellites enables a GPS receiver on Earth to establish location by communicating with at least three of the orbiting satellites within its "field of view." If contact can be made from a fourth satellite, altitude data can also be obtained. GPS is used in various industries including navigation (ocean and land), energy exploration, and environment monitoring. GPS is the essential component in automobiles equipped with navigation systems.

GSM
(wireless) Global System for Mobile Communications. The predominant worldwide standard for digital wireless service. GSM operates at either 900 or 1800 MHz and is used extensively in Europe and in more than 120 countries. GSM is a TDMA (Time Division Multiple Access) technology and allows a number of calls to take place on a single channel or frequency at the same time. For more information, contact GSM World at www.gsmworld.com.

hacker
(computers) a) A highly-skilled and clever computer programmer who writes programs in assembly or systems-level languages. The term refers to the programmer's "hacking away" at tedious computer code when writing complex programs. b) (popular usage) A mischievous person who seeks unauthorized entry into computer systems. Such entry can be benign (simply for the satisfaction of succeeding) or malicious (for the purpose of doing harm). *See* virus.

handoff
(wireless) The transfer of a wireless communication from one cell site to another without disconnection.

handset
(wireless) A small handheld device used to send and receive wireless communications. Also called a cell phone, cellular phone, wireless phone, mobile phone, and in Germany, a "handy."

HDML
(computers) Handheld Devices Markup Language. A computer language and coding system used to enable wireless devices such as cell phones, wireless pagers, and other handheld devices with small display screens to view information from the Internet. HDML is a version of HTML (HyperText Markup Language) and a subset of WAP (Wireless Application Protocol). *See* HTML, WAP.

hertz (Hz)
A measurement of electromagnetic energy expressed as the number of complete cycles per second of a wave-like radio signal. One cycle per second is 1 Hz, 60 per second is 60 Hz. 1,000 Hz is expressed as 1 KHz (kilohertz), 1,000,000 Hz is expressed as 1 MHz (megahertz), and 1,000,000,000 Hz as 1 GHz (gigahertz).

high frequency (HF)
That portion of the radio frequency spectrum between 3 and 30 MHz, characterized by wavelengths of 100 to 10 meters. High Frequency is also called shortwave and is used extensively for medium and long range terrestrial communications, including amateur radio, CB (Citizens' Band) radio, and single sideband radio. See *radio frequency.*

home coverage area
(wireless) In a cellular service plan, a designated geographic area within which the user is not assessed long-distance or roaming charges.

home page
(Internet) The first page viewed (main access page) on an Internet Web site. All other pages are accessed from the home page. The home page of a well-designed Web site enables the user to intuitively understand and navigate through the content of the site.

Hotmail
(Internet) A Web-based e-mail provider that enables users to send and receive messages from any computer connected to the Internet. Users get a special Hotmail e-mail address (_____@hotmail.com) that they can use from home, an office, or from an Internet café anywhere in the world. This is a favorite of travelers as the e-mail address is permanent and has an inexpensive annual charge. In any browser type in *Hotmail.com.*

HTML
(Internet) HyperText Markup Language. The computer language used for creating hyperlinked or hypertext documents on the World Wide Web. This computer language (also called a document format) enables users to view content in a Web browser and navigate within a Web site and to other Web sites by clicking hyperlinked text or graphics.

HTTP
(Internet) Acronym for HyperText Transfer Protocol. The communications protocol used by computers and Web browsers to connect to Web servers and move HTML files across the Internet. For example, to establish a connection with the World Trade Press Web server, one types http as a prefix to the Web domain name (http://www.worldtradepress.com) in the address field of a Web browser.

hyperlink
(Internet) A link between two text or graphic objects. Pointing with the computer's cursor and clicking a hyperlink will transfer you to a specified page, file (text, graphic, audio, or video), or Web site. Hyperlinks are generally identified in computer documents by underlined colored text (often blue) or by graphic buttons. Hyperlinks form the foundation for navigation on the World Wide Web. Hyperlink and hypertext are used interchangeably. See hypertext.

hypertext
(Internet) A single word or text passage on a Web page that has been hyperlinked to a related text occurrence, page, file (text, graphic, audio, or video), or Web site. Hypertext passages are often referred to as a "link" or "hyperlink." Hypertext links are generally identified in computer documents by underlined colored text (usually blue). Hypertext and hyperlink are used interchangeably. See hyperlink.

IE
(Internet) Acronym for Internet Explorer. *See* Internet Explorer.

interface
(computers) a) The connection of two or more computer devices such as a computer and a printer, or a keyboard and a computer. b) The connection between an individual computer, its operating system software, application software, and the user. c) (popular usage) To work with another individual or organization.

Internet

(computers/Internet) The world's largest, fastest-growing, and most important computer network. Specifically, a worldwide network made up of smaller networks that provides access to all those connected to any part of the network.

The original Internet was started in 1969 as the ARPAnet and was designed as a series of high-speed links between educational and research institutions. By the 1990s commercial and other traffic increased exponentially and the Internet became an economic and cultural phenomenon.

While the Internet's exact future is unclear, the following is generally understood to be true: 1) Internet use will continue to grow, 2) the Internet is by far the world's greatest source of information, and 3) the Internet holds untold opportunities for commerce and education. See World Wide Web.

Internet address

(Internet) a) An individual's e-mail address on the Internet. b) An individual Web site address (URL) on the Internet. *See* e-mail, Internet domain name, URL.

Internet café

A business that rents computers with Internet access by the hour and offers traditional "coffee house" food and drink services. The term is rather loosely used. In some cases it refers to a single old computer with a slow modem in the corner of a dusty general store. It can also refer to a huge operation with hundreds of state of the art computers with fast Internet access, comprehensive computer-related services (laser printers, scanners, etc.), plus full restaurant and full bar. Also called cyber café. There are now many thousands of Internet cafés in virtually every country of the world.

Internet discussion group

See newsgroup.

Internet domain name

(Internet) The address of an Internet Web site. Technically, an organization's domain name combined with a top-level domain (TLD). TLDs are extensions to the domain name such as .com, .org, .net, etc. For example: www.worldtrade-press.com. Note that no two organizations can hold the same Internet domain name and that trademarked names can only be used by the trademark holder. One can register an Internet domain name at any number of Web sites, but the most important register is Network Solutions (www.networksolutions.com). An Internet domain name is one of the components of a URL.

Internet Explorer

(Internet) The Web browser developed by Microsoft Corporation (www.microsoft.com), and referred to simply as Explorer or IE. Along with Netscape Navigator, Explorer is one of the two most popular Web browsers, although Explorer has been gaining market share and is now dominant. *See* Netscape Navigator.

Internet Service Provider

See ISP.

intranet

(Internet) A private computer network designed for use by one company or organization. For example, a company might design and maintain an intranet to share databases and provide intra-company e-mail services for its employees. An intranet operates on the same principles as the Internet. *See* Internet, extranet.

ISDN

(Internet) Integrated Services Digital Network. A technology that dramatically increases the data transmission capacity of telephone lines into homes and offices. With ISDN one can send analog and digital data over the same network at transmission rates of up to 128 Kbps.

Euro-ISDN, is an ISDN service that conforms to an agreed European norm to avoid incompatibilities between service and manufacturers; 26 network operators in 22 European countries signed an ISDN Memorandum of Understanding in 1989. *See* bandwidth, DSL, E1, E3, modem, T1-3.

ISP

(Internet) Internet Service Provider. A firm that provides clients with direct access to the Internet. ISPs may provide other services such as Web hosting and Web site design and building. An ISP provides a local telephone number for your computer to call to connect to the Internet. Online services such as AOL and CompuServe also provide Internet access as part of their service offering. For a directory of ISPs see Boardwatch Magazine's "Internet Service Providers" at www.boardwatch.com.

ITU

International Telecommunications Union. An agency of the United Nations with responsibilities for developing operational procedures and technical standards for the use of the radio frequency spectrum, telecommunications satellite orbits, and for the international public telephone and telegraph network. There are over 180 member nations of the ITU. The Radio Regulations that result from ITU conferences have treaty status and provide the principal guidelines for world telecommunications.

In the case of the U.S., they are the framework for development of the U.S. national frequency allocations and regulations. The ITU has four permanent organs: the General Secretariat, the International Frequency Registration Board (IFRB), the International Radio Consultative Committee (CCIR), and the International Telegraph and Telephone and Consultative Committee (ITTCC).

International Telecommunications Union
Place des Nations
CH-1211 Geneva 20, Switzerland
Tel: [41] (22) 730-5111
Fax: [41] (22) 733-7256
Web: www.itu.int.

Java

An object-oriented programming language developed by Sun Microsystems and designed to run on all computer operating systems, especially in Web applications.

A Java "applet" is a Java program that is first downloaded from a host Web server to a user's local computer, where it runs after being called from a Web page.

A Java "servlet" is a Java program that resides on and operates from a Web server.

As a "cross-platform" language, Java is distinguishable from a platform-specific language such as Microsoft's Visual Basic, which requires a Windows operating system for execution. The developers of Java used many valued features of the C and C++ languages, while eliminating much of the more arcane and difficult aspects of those languages. The Java language has become the most common first language in computer science curricula around the world.

JPG or JPEG

Joint Photographic Experts Group.

1) An International Standards Organization (ISO) group that develops and maintains standards for graphic file compression.

2) A still image (graphics) file compression format used extensively in displaying graphics on Web pages. The user has the ability to choose among a range of compression options that affect image quality and file size. The JPG compression format produces higher resolution images and more complex color palettes than the GIF format but results in a larger file size.

KB

See kilobyte.

Kbps

(computers/Internet) Acronym for kilo (1,000) bits per second. *See* bandwidth, bit.

keyword search

(Internet) A search for information or documents on the World Wide Web based on use of a single word, phrase, or combination of words in a search engine. By modifying the keywords and their sequence one can often obtain different search results.

kilobyte (KB)

(computers) One thousand bytes. *See* byte.

LAN
(computers) Acronym for Local Area Network. A computer network that is shared by computers and devices within a relatively small area. A LAN may serve two or three users or it may be used by hundreds, but the geographic area served ranges typically from a single office to part or all of a single floor of a building.

land line
(wireless) a) A traditional fixed location telephone. b) A telephone connection by wire or cable rather than by wireless.

laptop
(computers) A small, portable, battery-operated computer with integrated screen and keyboard.

LCD
Acronym for Liquid Crystal Display. A technology used extensively for displays in laptop computers, PDAs and other small computing devices. LCD displays use either a passive or active matrix display grid. Active matrix displays have a transistor at each pixel intersection point on the display, provide sharp images with good contrast and have a much faster refresh rate than passive matrix technology.

link
See hyperlink.

lithium ion
A lightweight and powerful rechargeable battery used in portable electronic devices such as wireless phones and laptop computers. Lithium ion batteries are more expensive than nickel cadmium (NiCad) batteries, but last longer.

local area network
See LAN.

locked, locked cell phone
(wireless) 1) A cell phone that requires a code to be entered on the keypad to be operational.
2) A GSM cellular phone that has been programmed to only accept a certain provider's SIM card.

Mac OS
(computers) Short for Macintosh Operating System. The operating system used on Macintosh computers. The Mac OS is especially popular in desktop publishing, design, and graphics-related industries.

MB
See megabyte.

Mbps
(computers) Acronym for mega (1,000,000) bits per second. See bandwidth, bit.

megabyte (MB)
(computers) One million bytes. *See* byte.

menu
(computers) An on-screen list of computer functions or operations that may be selected by the user.

MHz
Megahertz. See Hertz.

MMS
(wireless) Multimedia Messaging Services. A technology that allows cellular phone users to send and receive multimedia messages including text, still images, video and audio.

mobile e-commerce
(Internet) Electronic commerce that uses a wireless device as the access point (as opposed to a PC that is connected by landlines to the Internet). Examples of wireless devices used in mobile e-commerce include cellular telephones, beepers, and PDAs (Personal Digital Assistants) with Internet access.

mobile Web
(wireless) Access to the World Wide Web from a mobile wireless device such as a cellular phone, beeper, or PDA (Personal Digital Assistant). Mobile Web

promises to be a huge area for growth because of users' ability to gain instant access to information and mobile e-mail.

modem
(computers/Internet) Contraction of Modulator-Demodulator. An electronic hardware device that connects a computer communications port to a telephone line or TV cable and thence to other computers or computer networks including the Internet. Technically, a modem is a digital-to-analog and analog-to-digital device that converts a computer's digital pulses to audio frequencies that can be transmitted over a telephone line and vice-versa. A modem is only one way in which a user can connect to the Internet. Modern modems connect to the Internet at speeds of up to 56,000 bps, slower than ISDN, DSL and T1, T2, and T3 lines.

MPEG
Acronym for Motion Pictures Experts Group. An evolving series of standards for the compression of digital video and audio files for multimedia delivery. MPEG compression sacrifices some data quality in order to achieve a smaller file size.
MPEG-1 Designed primarily for video CDs. MPEG-1 layer 3 (also know as MP3) is used extensively for audio.
MPEG-2 Designed for use in digital TV broadcasting and DVD disks. This is high quality but produces larger files than MPEG-1.
MPEG-3 Designed originally for High Definition TV (HDTV), but was then merged with the MPEG-2 standard. MPEG-3 should not be confused with MP3 which is MPEG-1 audio layer 3 and used for compressing audio files.
MPEG-4 Designed for speech and video synthesis, computer visualization and artificial intelligence.
MPEG-21 Deals with larger multimedia issues and is in development.

national number
(telecommunications) A telephone number that is typically comprised of a subscriber's area code + local number.

Navigator
See Netscape Navigator.

Netscape Navigator
(Internet) The Web browser developed by Netscape Communications Corporation (www.netscape.com), now part of AOL Time Warner (www.aol.com). Along with Microsoft's Internet Explorer, Netscape is one of the two most popular Web browsers, although it has been losing worldwide market share to Internet Explorer, which since 1999 is the most widely used Web browser.

network
(computers) A group of linked computers. A network can link as few as two computers in a small office (a Local Area Network), several hundred or thousand computers in a company wide network (a Wide Area Network), or tens of millions of computers on the Internet. *See* extranet, Internet, intranet, LAN (Local Area Network), WAN (Wide Area Network).

newsgroup
(Internet) A message board on an Internet Web site or portal. Also known as an Internet discussion group. A newsgroup is generally topic-specific and starts with a single user posting a query or comment that is followed by comments and queries by other users. Newsgroups have become an extremely popular addition to Web sites that wish to foster communications among and with members or users. The accuracy of the posted information is often highly suspect or downright wrong. Posting on a newsgroup with a true name is a guarantee that one's e-mail address will be inundated with SPAM.

nickel cadmium (NiCad) battery
An early generation of rechargeable battery used in portable electronic devices such as wireless phones and laptop computers. NiCad batteries are being replaced by more modern, lighter weight, and longer-lasting batteries such as lithium ion, nickel metal hydride, and zinc air. Cadmium is toxic, which is one reason that NiCad batteries are being phased out.

nickel metal hydride (NiMH)
A powerful rechargeable battery used in portable electronic devices such as wireless phones and laptop computers. NiMH batteries are more expensive than nickel cadmium (NiCad) batteries, but last longer. They self-discharge faster than NiCad.

NMT
(wireless) Nordic Mobile Telephony. Scandinavian analog cellular phone technology from the 1980s, designed to work in uneven, hilly terrain. NMT 450 MHz and 900 MHz networks were installed in many parts of the world, but have been largely replaced by GSM cellular networks.

noise
(communications) Unwanted information in a signal, specifically a radio or wireless signal. Noise can be cross talk, hissing, popping, or any other unwanted sound in a wireless conversation.

offline
(Internet) Not connected to the Internet.
(computers) a) Describes a device (such as a laser printer) that is not connected to a computer. b) Describes a device that is not in ready mode.
(general) Not available for use.

off peak
(wireless) That time of the day or week when a wireless carrier experiences the least traffic. Off peak hours are usually before 7AM in the morning and after 7PM in the evening as well as Saturdays and Sundays. Air time rates are generally lower during off peak hours.

online
(Internet) Connected to the Internet.
(computers) a) Describes a device (such as a laser printer) that is connected to a computer. b) Describes a device that is in ready mode.
(general) Available for use.

online service
(Internet) A business entity that provides access to the Internet, proprietary databases and e-mail services to its users. Several of the largest online services include:
America Online (AOL) (www.aol.com)
CompuServe (www.compuserve.com)
DIALOG (online databases) (www.dialog.com)
Dow Jones Interactive (business, finance and news) (bis.dowjones.com)
LEXIS-NEXIS (legal and news information) (www.lexis-nexis.com)
Prodigy (www.prodigy.com)
WESTLAW (legal databases) (www.westpub.com)

operating system (OS)
(computers) The primary software program that manages a computer's basic functions and controls the execution of application programs. All application software must conform to certain standards of a computer's operating system to function. The most common operating systems are Windows (for Intel/IBM PC-compatible computers), Mac OS (for Macintosh), and UNIX and Linux (for PC and UNIX computers).

PACS
(wireless) Personal Access Communications System. A wireless telephone local area network system that acts like a cellular network, but works only in a short communications range of several hundred feet in each direction.

paging
(wireless) An optional wireless messaging service feature that transmits a voice or alphanumeric signal to the user

parallel port
A computer input/output data transfer interface that allows up to 8 bits of data to be transmitted at a time. Parallel ports use a 25-pin connector (DB-25) to connect external devices to a computer. This port is often labeled LPT1, and if the computer has additional parallel ports, LPT2, LTP3, etc. The most common use of a parallel port is to connect a printer to a computer. See also *serial port*.

PBX
(telecommunications) Private Branch Exchange. A subscriber-owned and located telephone network that is connected to outside lines and a carrier's telephone network.

PC
(computers) a) Personal Computer. Specifically a personal computer (laptop or desktop) running Windows or DOS. b) Any laptop or desktop personal computer. c) Acronym for printed circuit.

PCS
(wireless) Personal Communication Services. A second-generation digital wireless two-way voice, messaging, and data communications service operating at 1900 MHz. PCS is usually packaged with call waiting, voice mail, and caller ID service features.

PDA
(computers) Personal Digital Assistant. A handheld, battery-powered computing device used to store addresses, personal calendars, and notes. PDA and wireless technologies are merging to create combination PDA, wireless phone, and Internet-accessible devices.

PDF
Acronym for Portable Document Format. *See* Acrobat.

peak hours or peak time
(wireless) That time of day or week when a wireless carrier experiences the greatest traffic. Peak hours are usually Monday through Friday 7AM to 7PM. Rates are generally higher during peak hours.

Pentium
(computers) The world's dominant series of 32-bit CPU microprocessor chips manufactured by the Intel Corporation (www.intel.com) and used in PCs (personal computers). Pentium can refer to either the chip or a computer that uses the chip.

PIN
(wireless) Personal Identity Number. A code number used as a security password by the authorized owner or user for accessing a wireless phone, bank accounts, and other services and accounts.

pixel
(computers) Contraction of Picture (slang "pix") Element. The smallest element of visual information that can be used to make an image on a computer monitor or TV screen. The more pixels, the higher the resolution of a monitor. A pixel on a color monitor is made up of red, blue, and green dots of varying intensity that converge at the same point.

platform
(computers) a) A specific computer hardware architecture. For example, the PC, Macintosh, or UNIX platforms. b) A specific computer software architecture. For example, the Windows platform, or the Mac OS platform. The terms "platform," "operating system," and "environment" are often used interchangeably. *See* cross platform, operating system.

plug and play
(computers) Specialized computer software that recognizes when a new hardware component has been plugged into the computer system and that initializes and installs the component for immediate use with the system.

POP
(computers) Point of Presence. The point at which a telephone line or data connection from a user connects with a telephone company or an ISP (Internet Service Provider). In simple terms, this is the telephone number a dial-up modem calls to make an Internet connection from an ISP.

POP (POP3)
(computers) Post Office Protocol (Post Office Protocol 3). An Internet-based mail server that holds incoming e-mail until the user logs on and instructs it to be downloaded.

POP server
(computers) A server that implements the Post Office Protocol.

port

Any data pathway into or out of a computing or network device such as a computer, printer, scanner, router, switch or modem. For example, a modem is connected to the serial port on a computer. Also, every Ethernet ready device has a port for connecting it to the network.

portal

(Internet) A Web site that provides a variety of services to the user including Web search functionality, directories, news, e-mail, databases, discussion groups, online shopping, and links to other sites. The term was originally used to refer to all-purpose sites such as AOL and CompuServe, but is now more commonly used to refer to vertical market sites that cover specific industries or topics.

Predictive Text Input

(wireless) A widely used smart software on the latest GSM phones that assists in text messaging. The software predicts words before the user finishes entering them; this feature minimizes the number of key punches the user needs to make.

premium number

(telecommunications) A special services business telephone number that bills the caller an additional (usually per minute) fee in addition to the standard toll charge.

prepaid cellular/wireless

(wireless) A service plan by which a cellular user can pay in advance for airtime usage. Prepaid services are especially helpful for travelers and other persons who do not regularly subscribe to the locally available wireless service. They are also popular in countries where credit is not usually extended, and they provide an immediate means of tracking and controlling the cost of wireless services.

protocol prefix

See URL.

Quicktime

A multimedia (audio/video) development, storage and playback technology developed by Apple Computer. A Quicktime player enables the user to view a multimedia file which may contain audio, video, text and animation. Quicktime file extensions are *.qt, *.mov, *.moov.

radio frequency

That part of the electromagnetic spectrum in which electromagnetic waves can be generated by alternating current fed to an antenna. All wireless technologies, including television, AM and FM radio and WiFi are based on radio frequency field propagation. Radio frequencies are divided into the following parts of the electromagnetic spectrum:

RADIO FREQUENCY BANDS				
Band Name	**Abbr**	**ITU***	**Frequency**	**Wavelength**
Extremely low frequency	ELF	1	3–30 Hz	100,000 km – 10,000 km
Super low frequency	SLF	2	30–300 Hz	10,000 km – 1000 km
Ultra low frequency	ULF	3	300–3000 Hz	1000 km – 100 km
Very low frequency	VLF	4	3–30 kHz	100 km – 10 km
Low frequency	LF	5	30–300 kHz	10 km – 1 km

RADIO FREQUENCY BANDS				
Band Name	**Abbr**	**ITU***	**Frequency**	**Wavelength**
Medium frequency	MF	6	300–3000 kHz	1 km – 100 m
High frequency	HF	7	3–30 MHz	100 m – 10 m
Very high frequency	VHF	8	30–300 MHz	10 m – 1 m
Ultra high frequency	UHF	9	300–3000 MHz	1 m – 100 mm
Super high frequency	SHF	10	3–30 GHz	100 mm – 10 mm
Extremely high frequency	EHF	11	30–300 GHz	10 mm – 1 mm

*ITU Band, ITU = International Telecommunications Union.

RAM
(computers) Random Access Memory. Memory chips that plug into a computer's motherboard and that serve as the computer's primary but temporary workspace. Generally, the larger the files that are in use at one time, the more RAM a computer needs to operate. For example, a computer needs more RAM to process graphics files than textual word-processing files. Most new computers (as of 2002) are made with a standard minimum of 128 MB (megabytes) of RAM and allow for additional upgrades. Note that storage of data in RAM is temporary and is lost when the computer is turned off. See ROM (Read Only Memory).

Random Access Memory
See RAM.

Read Only Memory
See ROM.

re-boot
(computers) To restart (turn off, then turn on) a computer. This resets the computer, often resolving conflicts in the processing. See boot.

roaming
(wireless) Use of a wireless device outside the home service area. Most wireless service plans charge a higher per-minute rate for roaming calls.

roaming agreement
(wireless) An agreement between wireless carriers to allow their respective customers to use cellular phones or other wireless devices in each other's service area. Users are billed by their local carrier, which passes some of the revenue on to the roaming carrier.

ROM
(computers) Read Only Memory. Memory chips built into the motherboard of a computer that permanently store data, instructions, and routines for the operation of the computer. ROM cannot be altered by the user and is not lost when the computer is turned off. See RAM.

search engine
(Internet) Computer software designed to search the Internet for information based upon keywords entered by the user. Search engines do not reside on the user's computer, but rather at the online location of the respective company. Major search engines on the Web include:
AltaVista (www.altavista.com)
Excite (www.excite.com)
Google (www.google.com)

Hotbot (www.hotbot.com)
Infoseek (www.infoseek.com)
Lycos (www.lycos.com)
Northern Light (www.northernlight.com)
Overture (www.overture.com)
WebCrawler (www.webcrawler.com)
Yahoo! (www.yahoo.com)

SSL

Secure Sockets Layer. The leading security protocol on the Internet, originally designed by Netscape Communications. When SSL has been implemented on a Web site, the server sends a special key (code) to any browser requesting information from the site; the browser uses this key to send back a randomly-generated secret key, which is used to protect that client's session through encryption and authentication. The presence of SSL on a Web site is indicated by the prefix "https://" rather than "http://" in a Web site's URL.

serial port

A computer input/output data transfer interface in which one bit of data is transmitted at a time. A serial port is often used to connect a computer to a modem, PDA, printer or camera, but is a very slow interface. Parallel and USB ports are much faster. The serial port on a computer uses the RS-232 protocol and is often called the RS-232 port. Many latest-generation computers no longer provide serial ports, using the much faster and easier to use USB connections.

server

(Internet) A computer connected to a network so its data and programs can be shared by users of connected computers. Server can refer to either or both the hardware or the software that runs the computer itself. Some of the many types of servers include: Application Server (runs application software for multiple computers); Database Server (maintains a database accessed by multiple users); Remote Access Server (provides access to information to multiple remote users); and Web Server (provides Web access to the Internet).

service area

(wireless) The geographic area served by a wireless telecommunications carrier. The same as coverage area.

service plan

(wireless) A package of wireless services generally including system access, a number of peak minute usage, a number of off-peak minute usage, and provisions for long-distance, roaming, voice mail, Web access, and other optional services, based on a fixed monthly fee.

sideband

In radio communications, a band of radio frequencies either above or below the carrier frequency that contain energy as a result of modulation. In typical AM radio transmissions, both sidebands are present.
See also single-sideband.

single-sideband (SSB)

In radio communications, a form of amplitude modulation (AM) where the AM carrier and one of the sideband signals is removed. Single sideband radios are more efficient in their use of electrical power and bandwidth and are therefore effective in transmitting signals long distances.

SIM Card

(wireless) Subscriber Identification Module Card. A small plastic card embedded with a computer chip that fits into a GSM cell phone that contains a subscriber's personal information, phone settings (such as phone number), and a set amount of prepaid outgoing and incoming cell phone usage time. A GSM cell phone must have a SIM card inserted to become activated. SIM cards are transferable among cell phones. SIM Cards, also known as "smart cards," are transferable among GSM cell phones and are sold in most countries in telecom stores, news stands, and even from vending machines. The purchase of "prepaid" SIM cards has become enormously popular among GSM cell phone users as they allow the purchaser to initiate calls for a specific, cumulative number of minutes and to avoid monthly fees and other costs associated with standard subscription services.

slamming
(wireless) The unauthorized switching of a customer's telephone or wireless service from one carrier to another. Slamming is illegal in the U.S.

Smart Card
Popular name for SIM card. See SIM card.

SMS
(wireless) Short Messaging Service. Popularly known as "text messaging". The sending and receiving of short text messages between GSM cell phones (or between GSM cell phones and e-mail servers). Since text messages require little bandwidth and do not have to be sent in real-time, they are significantly less expensive than voice calls. Not surprisingly, SMS is the most popular cell phone communication method in the world today (in January 2003 1.65 billion text messages were sent in the UK alone).
The size limit on text messages has led to a worldwide, English-based text-messaging shorthand that lets users squeeze more input into each message (for example: 'r' for 'are' and 'u' for 'you'). See also EMS.

spam
(Internet) Unsolicited e-mail, usually mass e-mail messages promoting a political or social message or advertising a product. Named after the canned meat by-product SPAM (which is a registered trademark of Hormel Corporation). The use of mass e-mailings is the subject of great controversy. Some feel that unsolicited e-mail is a violation of the basic values of the Internet, while others have built legitimate businesses on such "mailings." See e-mail.

standby mode
(wireless) A cellular phone that has been turned on, but is not in use. This enables the user to receive incoming calls.

standby time
(wireless) The amount of time a fully charged cellular phone will operate in standby mode before fully discharging its battery. Each phone manufacturer states how long each model will operate in standby mode and/or in talk mode. Standby time is determined by the energy usage of the particular phone model as well as the type and size of the battery used.

subscriber
(telecommunications) A contractual user of a telecommunication carrier's cellular, wireless, satellite or land line phone service.

subscriber number
(telecommunications) The telephone number of a telecommunications subscriber.

T1, T2, T3
(Internet) Dedicated digital circuit connections that are leased from a telecommunications provider and that provide 1.544 Mbps (T1), 3.152 Mbps (T2), and 44.736 Mbps (T3) data transmission capacity. T1-3 lines are used for high speed private networks and connections to the Internet. T3 lines can handle full-screen, full-motion video. The European equivalent is E1-3. *See* bandwidth, DSL, E1, E3, ISDN, modem.

TACS
(wireless) Total Access Communications System. A wireless technology based on analog AMPS technology, but operating in the 900 MHz frequency and used primarily in the UK. TACS has been largely replaced by GSM, although there are still TACS networks in some British Commonwealth countries. There has been one enhancement called ETACS (Extended TACS).

talk time
(wireless) The amount of time a fully charged cellular phone will operate while in talk mode before fully discharging its battery. Each phone manufacturer will state how long each model will operate in standby mode and/or in talk mode. Talk time is determined by the energy usage of the particular phone model as well as the type and size of the battery used. Talk time uses more battery life than standby mode.

tax impulsing (tax metering)
(telecommunications) A high-frequency signal (pulse) used by telecommunications carriers in some European countries to meter telephone calls and establish how much to bill the subscriber. Tax impulsing often interferes with data transmissions. See "Tax Impulsing is Disconnecting My Modem" on page 143.

TCP/IP
(Internet) Transmission Control Protocol / Internet Protocol. The most fundamental communications protocols of the Internet. A set of procedures designed to allow computers to share resources across networks. A community of researchers developed these protocols in the early 1980s while working on the original ARPAnet (predecessor to the Internet). The IP protocol tags data with an address in much the same way as addressing and sending a envelope to its destination. The TCP protocol splits data into packets (think envelopes) for efficient transmission, and then reassembles these packets on the receiving end. The whole Internet is based on TCP/IP.

TDMA
(wireless) Time Division Multiple Access. An advanced digital wireless communications technology operating at 900 MHz that allows many users to share a single radio frequency without interference.

toll charges
(wireless) A charge for making a long-distance call outside of the user's service area.

toll-free number
(telecommunications) A special service telephone number that is free to the caller.

toll-free calling area
(wireless) A geographic area within which a user may make calls without paying long distance or toll charges.

top-level domain
(Internet) The primary organizational category of a domain name. All domain names are organized into categories that are assigned and administered by ICANN (Internet Corporation for Assigned Names and Numbers). The top-level domain is indicated by the extension at the end of a domain name and is intended to describe the nature of the domain name owner. Top-level domains include:

.com —commercial (business entities)
.net —network
.org —organization (non-government organization)
.edu —U.S. educational institution
.gov —U.S. government (federal, state, or local)
.mil —U.S. military

The exponential surge in domain name registrations has led to the introduction of new top-level domains, some of which include:

.biz —business entity
.pro —professional
.museum —museum
.info —information service
.name —individual person

Many countries are permitted to have top-level domains, indicated by two-letter extensions. A few examples include:

.ca —Canada
.fr —France
.uk —United Kingdom

tri-band (mode) phone
(wireless) A mobile phone (handset) that supports GSM 900/1800/1900 MHz bands, or alternatively 850/1800/1900 bands. Different countries use different frequency bands for mobile networks. A tri-band phone is designed to operate on three of these bands, favoring two of the home country's bands and one for roaming in other countries.

trunk
(telecommunications) A circuit between two telephone exchanges; essentially a long-distance telephone line.

UHF

Acronym for Ultra High Frequency. That portion of the radio frequency spectrum between 300 MHz and 3 GHz and characterized by wavelengths of 1 m to 100 mm. The most common uses of UHF are for television and cellular phones and related devices. Since UHF radio waves are very short compared to VHF and lower frequencies, the receiving and transmitting antennas can likewise be much smaller. See radio frequency.

UM

Unified Messaging. The integration of different communications media such that voice, fax and e-mail messages may be sent and received from a single interface, such as a landline phone, wireless phone or personal computer.

UMTS

(wireless) Universal Mobile Telecommunications Standard. Third-generation (3G) European wireless standard based on W-CDMA and using the 2GHz bandwidth. UMTS is intended to be a worldwide standard for European GSM networks of the future. The UMTS is an ambitious standard intended to build upon and enhance wireless technologies through greater data transmission rates, increased capacity and a new range of wireless services.

up

(computers/Internet) a) A computer that has been turned on or that is in operational condition. b) A Web site that is operational. *See* down.

upload

(Internet/computers) The transfer of computer documents or files from a local (client) computer to a remote computer over a network. To download means to receive and to upload means to transmit. See download.

URL

(Internet) Acronym for Uniform Resource Locator. The address and/or route to a file, document, or Web site on the Internet. A URL is typed into the "address" window of a Web browser to access a particular Web domain or Web page on the Internet. For example:

http://www.worldtradepress.com is the URL for World Trade Press.

Specifically, a URL is a series of letters, characters, and numbers that contain 1) a protocol prefix, 2) port number, 3) domain name, 4) subdirectory name, and 5) file name. The port number is almost always a default and does not need to be included. As a result, a user need only type in the protocol prefix and domain name to gain access to a Web site or page. There are more than 10 protocol prefixes, but the most common are:

http: —World Wide Web server
ftp: —file transfer protocol server
news: —Usenet newsgroups
mailto: —e-mail
gopher: —Gopher service
file: —file on local system

In many cases a specific page is stored in a subdirectory on the domain. For example:

http://www.worldtradepress.com/catalog/dictionary.html is the specific Web page for the World Trade Press Dictionary of International Trade. This URL can be broken up as follows:

http: —protocol prefix
// —separators
www.worldtradepress.com/ —domain name
catalog/ —subdirectory name
dictionary.html —document name

See Internet domain name, top-level domain.

USB

Acronym for Universal Serial Bus. A universal interface standard for communication between a computer and an external peripheral device using bi-serial transmission. In practical terms, a "plug-and-play" computer cable that enables the user to easily connect a computer to external devices such as keyboards, printers, scanners, modems, joysticks etc. without having to turn the computer off. The USB standard 1.x supports data transfer rates of 12 Mbps (12 million bits per second) and USB 2.0 supports data transfer rates of 480 Mbps.

VHF
Acronym for Very High Frequency. That portion of the radio frequency spectrum between 30 MHz and 300 MHz, characterized by wavelengths of 10 m to 1 m and used for short range communications. The most common uses of VHF are FM broadcast (88-108 MHz range) and television broadcast (also uses UHF). See radio frequency.

virus
(computers) A software program designed maliciously to infect and harm computer files. A computer virus is distinguished by its ability to attach itself to other programs and self-replicate. Typically, a virus arrives hidden within or attached to a file, program, e-mail, or e-mail attachment. Some viruses simply display a harmless message when activated, while others are designed to do severe damage to computer files and system software. A virus can be activated automatically or by a trigger event such as a date, time, or sequence of keyboard strokes. A number of anti-virus programs are available and can be especially helpful because many viruses operate on similar principles that can be detected by the anti-virus software. However, it is wise to regularly update this software (often with direct downloads from Web sites) as new viruses are invented and discovered daily. See hacker.

voice-activated dialing
(wireless) A wireless telephone feature where the user speaks keywords into the handset and a specific telephone number is automatically dialed. Examples of keywords are: "Call Home," "Husband," and "Office."

voice mail
(wireless) A telephone service feature that answers a user's calls when the user is not available, plays a greeting often in the user's own voice, records the caller's message, and then enables the user to review, save, delete, or reply to the recorded message.

WAN
(computers) Wide Area Network. A data communications network that serves a very large geographic area such as a state, province, or country. *See* extranet, Internet, intranet, LAN (Local Area Network), network.

WAP
(computers/Internet) Wireless Application Protocol. A set of communications protocols designed to provide e-mail, Internet Relay Chat (IRC), and Web content to wireless devices such as cellular phones, beepers, and wireless PDAs (Personal Digital Assistants). The development of WAP was initially spurred by four major manufacturers of wireless products: Motorola, Nokia, Ericsson, and Unwired Planet. WAP technology is based on WML (Wireless Markup Language), which is a derivative of HDML and HTML designed for small screen displays. See HDML, HTML.

W-CDMA
(wireless) Wideband CDMA. (European version of Qualcomm's CDMA and the 3G evolutionary step planned for GSM.) Also referred to as B-CDMA (Broadband CDMA), but not to be confused with the similar CDMA2000. W-CDMA is a considerable enhancement of the original Qualcomm CDMA and is an integral part of the 3G (Third Generation) UMTS standard. W-CDMA has 10 times the computational capacity of 2G technologies.

Web browser
(Internet) An application program that enables the user to access the World Wide Web on the Internet. To view a particular Web site or page, the user types its Internet address (URL) into the "Address" window of the Web browser. Today's popular browsers enable the user to view graphics and text as well as listen to audio files and see video files that reside on the World Wide Web. Technically a browser is a client application that uses the HyperText Transfer Protocol (HTTP) to enable the user to make requests of Web servers. Popular browsers include Microsoft Internet Explorer and Netscape Navigator, both of which support e-mail. See Internet Explorer, Netscape Navigator.

Web page
(Internet) A document on the World Wide Web. A Web page is maintained on a Web site. *See* Web site, home page.

Web server
(computers/Internet) A computer with an operating system and server software designed to provide World Wide Web (WWW) services on the Internet. After an organization's Web site data is loaded onto its Web server, it is accessible to anyone accessing the WWW. A Web server can be located in an organization's offices or at a third-party hosting service.

Web site
(Internet) A collection of files on a server that are accessible on the World Wide Web. The term is used both to denote the server that contains files and the collection of files that relate to a specific URL. A Web site is composed of a "home page" and other data files that can be accessed from the home page.

wireless
(wireless) The use of radio frequencies and equipment to transmit and receive voice, data, and video communications.

wireless Internet
(wireless) Service, equipment, and software that enables a wireless device user to access the Internet.

WML
Wireless Markup Language. A tag-based computer language used especially for coding content for transmission to, display on, and transmission from wireless devices such as cellular phones and PDAs. Like its competitor cHTML (Compact HTML), WML aims at the minimum necessary set of tags, due to the limited memory and storage capacity of wireless devices. WML was originally developed for use in the Wireless Application Protocol (WAP).

World Wide Web (WWW) (Web)
(Internet) a) The aggregation of documents that can be accessed via the Internet. b) The aggregation of hypertext servers (HTTP servers) that can be accessed via the Internet. c) The system of accessing files and documents on the Internet using hypertext links.

The Web operates using the HTTP protocol and provides graphic capability. The Internet, on the other hand, is the network of computer networks and servers on which the Web resides and operates. The Web was developed at the European Center for Nuclear Research (CERN) between 1989 and 1991 as a means of sharing data on nuclear research.

WWW
See World Wide Web.

XML
eXtensible Markup Language. An open standard computer language used for defining data elements, most commonly on Web pages. XML is considered the successor language to HTML (HyperText Markup Language) because of its ability to define content and function as if it were a database. In HTML the Web developer defines how text and graphic elements are displayed, whereas in XML the developer defines what these text and graphic elements contain. In HTML, the developer can use only predefined tags, whereas in XML the developer can create and define the tags. Although XML freely lets developers create their own tags, there has been notable and continuing efforts in various industries to develop industry-wide, agreed-upon XML tags.

zip
(computers/Internet) A popular method of compressing a computer file or group of files. Compression of large documents makes them easier to store or send via e-mail as an attachment. After being "zipped" a file has the suffix ".zip" appended to its name. When extracted, the file returns to its normal size and can be opened using the original application program. Most computers and browsers come equipped with a zip utility and *.zip files can be extracted (uncompressed) simply by double-clicking the document icon or name.

Trip Notes

Destinations   Dates

1. _____   _____

2. _____   _____

3. _____   _____

Cell Phone Details
<u>Use Domestic Phone?</u>   <u>Details /SIM Cards?</u>   _____

Destination 1. _____   _____

Destination 2. _____   _____

Destination 3. _____   _____

Internet Connectivity Details
<u>Hotel 1</u>

Digital or Analog line: _____ Phone plug: _____ Electric plug: _____

Line Filter: _____ Notes: _____

<u>Hotel 2</u>

Digital or Analog line: _____ Phone plug: _____ Electric plug: _____

Line Filter: _____ Notes: _____

<u>Hotel 3</u>

Digital or Analog line: _____ Phone plug: _____ Electric plug: _____

Line Filter: _____ Notes: _____

Internet Cafés
Destination 1: _____

Destination 2: _____

Destination 3: _____

Special Phone Numbers
Destination 1: _____

_____

Destination 2: _____

_____

Destination 3: _____

_____

Notes
Destination 1: _____

_____

Destination 2: _____

_____

Destination 3: _____

_____

Notes